GAME THEORY

A Nontechnical

Introduction

GAME
THEORY
A Nontechnical Introduction

MORTON D. DAVIS
WITH A FOREWORD BY OSKAR MORGENSTERN

Basic Books, Inc., Publishers

NEW YORK / LONDON

To my wife, Gloria, who so painstakingly read and reread my manuscript and to my two- and four-year-old children, Jeanne and Joshua, who spent their infancy during the writing and who, whenever errors were uncovered, were always prepared to eat my words.

© 1970 by Morton D. Davis
Library of Congress Catalog Card Number: 79-94295
SBN 465-02626-5
Manufactured in the United States of America
Designed by Vincent Torre

Foreword

Oskar Morgenstern

Game theory is a new discipline that has aroused much interest because of its novel mathematical properties and its many applications to social, economic and political problems. The theory is in a state of active development. It has begun to affect the social sciences over a broad spectrum. The reason that applications are becoming more numerous and are dealing with highly significant problems encountered by social scientists is due to the fact that the mathematical structure of the theory differs profoundly from previous attempts to provide mathematical foundations of social phenomena. These earlier efforts were oriented on the physical sciences and inspired by the tremendous success these have had over the centuries. Yet social phenomena are different: men are acting sometimes against each other, sometimes cooperatively with each other; they have different degrees of information about each other, their aspirations lead them to conflict or cooperation. Inanimate nature shows none of these traits. Atoms, molecules, stars may coagulate, collide, and explode but they do not fight each other; nor do they collaborate. Consequently, it was dubious that the methods and concepts developed for the physical sciences would succeed in being applied to social problems.

The foundations of game theory were laid by John von Neumann, who in 1928 proved the basic minimax theorem, and with the publication in 1944 of the *Theory of Games and Economic Behavior* the field was established. It was shown that social events can best be described by models taken from suitable games of strategy. These games in turn are amenable to thorough mathematical analysis.

In studying the social world we are in need of rigorous concepts. We must give precision to such terms as utility, information, optimal behavior, strategy, payoff, equilibrium, bargaining, and many more. The theory of games of strategy develops rigorous notions for all of these and thus enables us to examine the bewildering complexity of society in an entirely new light. Without such precise concepts we could never hope to lift the discussion from a purely verbal state and we would forever be restricted to a very limited understanding if, indeed, we could achieve it at all.

It may appear that the mathematical theory remains inaccessible to the mathematically less advanced reader. But this is not so: it is possible to give a clear, comprehensive and penetrating account of the theory and of many of its applications if an important proviso has been fulfilled. He who attempts to do this, who wishes to give a verbalization of a higher order, must himself have a profound insight into all the intricacies of the theory and, if possible, should have participated in its development. These conditions are amply fulfilled by Morton Davis, the author of this admirable book. A new branch of science is, indeed, fortunate to have found an expositor of his caliber who can bring so many new and in some ways perplexing ideas to the general reader.

A book such as the present is in the best tradition of science writing that has long since been familiar in the natural sciences. In those fields—mathematics not exempted—very able writers have made great efforts to explain accurately but in as simple terms as possible the new results which have come with often bewildering rapidity. In fact, one is probably justified in assuming that these writings have themselves in turn contributed toward further advances since they have spread the knowledge and interest, and many a new mind has been attracted to the respective fields. In the social sciences books such as the present one are rare. This is partly due to the fact that there were few theories to report about comparable in scope and difficulty of those normally encountered in the physical sciences or in that respect comparable to game theory. Partly it was also difficult to avoid the injection of personal value judgments in the discussion of emotionally laden social and economic problems. The present book is entirely free from such faults; it explains, it analyzes and it offers precepts to those who want to take them; but the theory it describes and develops is neutral on every account.

The reader of this book will be impressed by the immense complexity of the social world and see for himself how complicated ultimately a theory will be that explains it—a theory compared to which even the present-day difficult theories in the physical sciences will pale.

The reader who will follow Davis on his path will be led into a new world of many beautiful vistas; he will see many peaks that have been conquered, but also many still calling for exploration. He will emerge from his trip, I am sure, with a better understanding of the intricacies of our life.

Princeton University
November 1969

Author's Introduction

*We hope, however, to obtain a real under-
standing of the problem of exchange by
studying it from an altogether different angle;
this is, from the perspective of a "game of
strategy."*—Von Neumann and Morgenstern,
The Theory of Games and Economic Behavior

The theory of games was originally created to provide a new
approach to economic problems. John von Neumann and
Oskar Morgenstern did not construct a theory simply as an
appendage, to take its place on the periphery of economics
as an afterthought. Quite the contrary. They felt that "the
typical problems of economic behavior become strictly identical with the mathematical notions of suitable games of
strategy." The term "game" might suggest that the subject is
narrow and frivolous, but this is far from the case. Since
the classical work of von Neumann and Morgenstern was
published, the theory of games has proven to be of sufficient
interest to justify its study as an independent discipline.
And the applications are not limited to economics; the effects of the theory have been felt in political science, pure
mathematics, psychology, sociology, marketing, finance, and
warfare.

Actually, the theory of games is not one theory but many.
This is not surprising. A game, after all, is only a model of
reality and it would be too much to expect that a single
model could accurately reflect such diversified types of

activity. But there are certain elements that are contained in all these models, and it is these that we will be concerned with when we discuss games.

The word "game" has a different meaning for the layman and for the game-theorist, but there are some similarities. In both types of games there are players, and the players must act, or make certain decisions. And as a result of the behavior of the players—and, possibly, chance—there is a certain outcome: a reward or punishment for each of the participating players.

The word "player," incidentally, does not have quite the meaning one would expect. A player need not be one person; he may be a team, a corporation, a nation. It is useful to regard any group of individuals who have identical interests with respect to the game as a single player.

How to Analyze a Game

Two basic questions must be answered about any game:
How should the players behave?
What should be the ultimate outcome of the game?

The answer to one of these questions, or to both, is sometimes called the solution of the game, but the term "solution" does not have a universal meaning in game theory; in different contexts it has different meanings. This is true also of the word "rational" when used to describe behavior.

These questions lead to others. What is the "power" of a player? That is, to what extent can an individual determine the outcome of a game? More specifically, what is the minimum a player can assure himself, relying on his own resources, if he receives no cooperation from others? And is it reasonable to suppose that the other players will in fact be hostile?

In order to answer these questions, one must know certain things about the game. First, one must know the rules of the game. These include:

1. To what extent the players can communicate with one another

2. Whether the players can or cannot enter into binding agreements

3. Whether rewards obtained in the game may be shared with other players (that is, whether side payments are permitted)

4. What the formal, causal relation between the actions of the players and the outcome of the game is (that is, the payoff matrix)

5. What information is available to the players

In addition, the personalities of the players, their subjective preferences, the mores of society (that is, what the players believe to be a "fair outcome"), all have an effect on the outcome.

Perhaps the single most significant characteristic of a game is the number of players, that is, its size; in fact, the games will be discussed in this book in order of size. Games of the same size will be grouped together and then distinguished from one another on the basis of other, less prominent differences. The smallest games will be discussed first.

Generally, the fewer the players, the simpler the game. As one progresses from the simplest games to those of greater complexity, the theories become less satisfying. This might almost be expressed in the form of a perverse, quasi-conservation law: the greater the significance of a game—that is, the more applications it has to real problems—the more difficult it is to treat analytically. The most satisfying theory, at least from the point of view of the mathematician, is that of the competitive, two-person game; but in real life the purely competitive game is rare indeed. For the more common, partly competitive, partly cooperative, two-person game, in fact, no generally accepted theory exists. This perverse law, incidentally, is not restricted to game theory alone. In the area of psychology, it is considerably easier to determine how rats learn than to discover the causes of mental illness.

In the more complex games, a player is faced with forces that he cannot control. The less control a player has upon the final outcome, the more difficult it becomes to define rational behavior. What constitutes a wise decision when any decision will have little influence on what ultimately happens?

Earlier we asked what a player *should* do and what the ultimate outcome of a game *should* be—"should," not in the moral sense, but in the sense of what will most further a player's interests. The question of what people *actually* do and what *actually* happens when a game is played is best left to the behavioral scientist. (This is sometimes expressed by saying that the game-theorist is interested in the normative rather than the descriptive aspects of a game.) This is all very well in the simpler games, but in complex games the distinction becomes blurred. If a player is not in control, he must be concerned with what the other players want and what they intend to do. If, for example, in a labor-management dispute, one party knows the other party has an aversion to risk, he might act more aggressively than he would otherwise. A short-term speculator on the stock market is almost wholly in the hands of others; what he *should* do depends on what he thinks others *will* do.

As the games grow more complex, it becomes almost impossible to give convincing answers to our two initial questions; we have to be content with less. Instead of determining a precise outcome, we often have to settle for a set of possible outcomes that seem to be more plausible than the rest. These possible outcomes may be more firm (we say stable) than the others in that no player or group of players has both the power and motivation to replace them with one of the less favored outcomes. Or the outcomes may be fair or enforceable in a sense that we will define later. Or, alternatively, we may take an entirely different approach and try to determine the average payoff a player would receive if he played the game many times.

New York City/December 1969   MORTON D. DAVIS

Acknowledgments

In the course of writing this book and in the years immediately preceding, I have profited from my contact with many others. From so many others, in fact, that I could not hope to acknowledge them all. Nevertheless, I would like to express my gratitude to those to whom I am particularly indebted. To Professor David Blackwell, my mentor at the University of California, who first introduced me to game theory; to Professor Oskar Morgenstern, who directed the Econometric Research Group at Princeton University where I worked for two years and who allowed me (and many others) access to an intensely stimulating, scholarly environment; to the Carnegie Corporation for financial support during these years; and to the many people with whom I had innumerable conversations—especially Professors Michael Maschler and Robert Aumann of the Hebrew University of Jerusalem.

Contents

GAME THEORY

A Nontechnical

Introduction

1

The One-Person Game

One-person games are by far the simplest type, at least conceptually—so simple, in fact, that some do not consider them games at all. In one-person games, it is convenient to regard nature as a person. Thus, one-person games are really games against nature. Nature differs from man, of course, in that you cannot anticipate its action by guessing its intentions, since it is completely disinterested.

One-person games may be grouped in three categories, on the basis of the role of nature in each. In the first category, nature has no active role whatever. The player simply makes a choice, and that choice determines what happens. In the second category, the laws of chance do take a part. The player chooses first, and then the rules of chance determine the rest. Here, the player is aware of the relevant probabilities in advance. In the third category, the player makes his decision without advance information on how nature will "play."

A person entering an elevator provides a simple example of the first category. His alternatives are the several buttons he can push. The possible outcomes are the different floors he can reach. And there is an obvious causal connection between one and the other. It is clear that one cannot sensibly "choose" a good strategy (that is, decide which button to push) until he decides where he wants to go. This is an important, if obvious, point and we will return to it later, when we discuss the concept of utility.

In principle, any problem in the first category is this simple; but, in practice, there may be complications. It may be possible, for example, for an expert to evaluate how profitable a chemical plant may be, once he is told its specifications. Yet it may be quite difficult to design the best possible plant, because there are so many possibilities.

As an illustration of the second category, consider the following example. Five suspected criminals are meeting secretly in the cellar of a building. Outside, a policeman with orders to follow the leader of the gang is waiting for them to disperse. The policeman knows that his man is the tallest of the group, and that is his sole means of distinguishing him from the rest. As a security measure, the men leave the meeting one at a time. The interval between successive departures is so long that if the policeman waits to see who leaves next before following any one of them, he will lose his opportunity. If the criminals leave the meeting in a random order, what is the best strategy for the policeman? If he adopts the best strategy, what is the probability that the person he follows will indeed be the leader?

At first glance, this may seem to be a many-person game; certainly, more than one person has an interest in the outcome. But since the suspected criminals are unaware of their immediate danger and are making no overt decisions concerning it, the game is really one of man against nature.

Were it otherwise—if, for instance, the criminals knew of the policeman's intentions and were conspiring to frustrate them—we would have a very different problem.

Since nature has no "interest" in the outcome, we may assume that any one particular order of departure is as likely to occur as any other. This being the case, the policeman's best strategy is to let the first two criminals go, and then follow the first after that who is taller than the ones who have preceded him. If, after all the criminals have gone, no such person has turned up, the leader has escaped. It can be shown, but we will not take the trouble to show it here, that the policeman will follow the right man about 40 percent of the time if he acts along these lines.

Surprisingly, the policeman can pick out the leader about 40 percent of the time, no matter how many people are in the cellar, as long as he knows in advance exactly how many that is. Just as before, he lets the first 40 percent (approximately) go by and follows the first person among the remaining 60 percent who is taller than the ones who have gone before.

It is not difficult to think of other circumstances in which decisions must be made on the basis of imperfect or incomplete information. A farmer who must decide whether or not to buy artificial irrigation equipment may not know the exact probability of rain but can make a reasonable estimate on the basis of past experience. An insurance company that has issued many fire-insurance policies within a small area can fairly well estimate the chances of going bankrupt and can reinsure or not with another company.

What contribution can the game-theorist make in this kind of problem? Almost none. The decision whether to gamble, and which gamble to take, is the player's; a preference for or aversion to gambling is a matter of taste. The player here is in much the same position as the passenger in an elevator choosing his floor.

While it is good to know the possible, or at least the probable, consequences of one's actions, it is, unfortunately, a fact that decisions often must be made in ignorance thereof. These situations constitute the third category of one-person games and are much more difficult to cope with. Consider this example:

Two competing manufacturers of buggies are offered the rights to a still imperfect invention: the horseless carriage. The first company accepts the offer and suggests that the two share it jointly. The second company must decide what to do. It can accept the offer, and take the loss if the invention fails or reap the profits if it succeeds. Or it can reject the offer, and suffer a substantial competitive disadvantage if the invention succeeds or break even if it fails. The key question is the likelihood of success. But there isn't any data on which to base even an educated guess: no one has ever made horseless carriages before.

Still another example: A serum is developed as a result of experimentation with animals. It is not clear whether it will be effective on humans, or whether there will be side effects. Yet a decision must be made to use it on humans or not.

In this type of problem, the game-theorist can contribute a little more. The basic philosophy must still be determined by the player: he can choose to look at the problem optimistically or he can expect the worst. But once the player makes his view clear, the game-theorist can select the tactics.

Suppose a player wants to play as conservatively as possible; if anything can go wrong, he feels, it will. The game-theorist would not question that attitude; he would simply assume that nature here is behaving like an antagonistic human being, and he would recommend what is called the minimax strategy: the suggested strategy for two-person, competitive games.

But is it reasonable to attribute malice to an indifferent nature? Does it make sense to assume that nature is out to get you? Admittedly, this is not a compelling assumption, but it is plausible as a guide for action, and not demonstrably worse than any other. In the area of statistical decision-making, an important type of one-person game, this is the most prevalent point of view.

The one-person game is relatively unimportant and we will not dwell on it. The essential point is that one must ascertain a player's goals before one can suggest a course of action.

2

Finite, Two-Person, Zero-Sum Games of Perfect Information

No one game can serve as a universal model; a single game is inadequate even to represent one group of games. We are, therefore, dividing finite, two-person games into three categories:

Zero-sum games of perfect information

General zero-sum games

Non-zero-sum games

In this chapter we will be concerned with zero-sum games of perfect information.

Rather than define terms and discuss the games abstractly, we will look at some examples and use these as the focal point for our discussion. It is not chance, incidentally, that we have selected an ordinary parlor game, chess, as an example; games of perfect information are almost exclusively parlor games. The converse is not true, however. Bridge and poker, for instance, are not games of perfect information.

Two chess players are scheduled for a match, but neither is able to appear at the time set. A substantial amount is at stake and, to avoid forfeiture, each player must submit to the referee, in writing, a description of how he proposes to play. The referee agrees to act as agent for both players and to move the pieces as directed, but he refuses to use his own judgment in making moves. The plans submitted must be so detailed, therefore, that they allow for all contingencies.

We naturally suppose that each player wants to win or, barring that, prefers a draw to a loss. We also assume that each player has the time and ability to look as far ahead as he wishes, and the time and space to write his plan in sufficient detail.

In such a game what is meant by a "good" strategy? Does a good strategy necessarily exist? Also, if a player has the opportunity to see his opponent's plan before submitting his own, how important is this advantage?

The purpose of analyzing a game is to solve simultaneously many apparently diverse problems, problems with basic common elements. The first step in moving from the concrete problem to the abstract game is to differentiate what is essential from what is not. To the eye of the game-theorist, there are four essential elements in the sample chess game: (1) there are two players; (2) they have opposite interests with respect to the outcome of the game*;

---

* Games in which the interests of the players are diametrically opposed are called zero-sum games. The term "zero-sum" is derived from parlor games such as poker where wealth is neither created nor destroyed. There, if you want to make money, you must win it from another player. After the game is over, the sum of the winnings is always zero (a loss is a negative win). This game is distinguished from a non-zero-sum, union-management bargaining game, for example, in that, there, both players may lose simultaneously if there is no agreement. That is, one man's loss may not be another's gain.

(3) the game is finite*; and (4) there are no surprises: at every point of the game, both players are completely informed. Games with these four properties are said to be two-person, zero-sum, finite games of perfect information. Not taken into account in this model are the size of the board, the number of pieces and the manner in which they move, the specific rules that determine who wins at the end, and other such particular points.

The advantage of studying categories of games by abstracting certain of their properties is that one can draw inferences that have general application. But there are disadvantages. The "flavor" of the game vanishes as the distinction between games disappears. Games of perfect information vary considerably; among them are chess, checkers, tick-tack-toe, and the Japanese game go. For our purposes, however, these games are identical, and any conclusions we draw must hold for all of them. Yet it is well known that experts in each of these games have very specialized knowledge; a good chess player may be only a beginner at checkers. Thus, it is reasonable to expect that our approach will not be of much value to the practical player.

Describing a Game

Any game can be described in more than one way; it will best suit our purpose to use what is called the "normal" form. Before explaining what we mean by the "normal" form of a game, let us first explain what we mean by strategy.

The concept of *strategy* is basic in game theory. A strategy is a complete description of how one will behave under every possible circumstance; it has no connotation of cleverness. In the game of tick-tack-toe, one strategy

---

* That is, a player has only a finite number of alternatives at each turn, and the game is terminated in a finite number of moves.

might be, at each turn, to play in the column as far to the right as possible, and of the unoccupied squares remaining in this column, to choose the highest one possible.

It follows, from the assumptions we made earlier, that the number of strategies possible to a player is finite, and so, in principle, they can be enumerated. In our sample chess game, the strategies for white will be denoted by S_1, S_2, S_3, etc., and those for black by T_1, T_2, T_3, etc. Since a strategy describes what will be done in any position that can come up, if you know each player's strategy you can predict the outcome of the game.

Suppose that, before the game starts, the players prepare the matrix shown below. This is always possible, since a player knows all of his opponent's possible strategies as well as his own. The strategies for white are listed vertically on the left, and the strategies for black are listed horizontally across the top. Inside the square that marks the intersection of a column corresponding to a strategy for black and a row corresponding to a strategy for white is the letter W, D, or L. By convention, this means that if the players use the indicated strategies the result will be a win, draw, or loss for white, respectively. In the matrix below, if black chooses T_3 and white chooses S_4, white will lose.

A game is said to be in *normal form* when the entire sequence of decisions which must be made throughout the game are lumped together in one, single decision: the choice of a strategy. In a real game such as checkers or chess this lumping process is purely conceptual. In practice the game is actually played in *extensive form* that is, the decisions are actually made one at a time.

The normal form is a particularly simple means of describing a game and for this reason it is of theoretical importance. In Figure 1 we see how a normal form description might look; white's strategies are listed vertically on the left, black's strategies are listed horizontally above; and within the matrix, for each pair of strategies for black and white, are listed the corresponding outcomes.

Strategies for Black

		T_1	T_2	T_3	T_4
Strategies	S_1	W	D	L	L
for	S_2	D	W	W	L
White	S_3	L	L	W	W
	S_4	D	D	L	W

Figure 1

Having described the game in normal form, we are now in a position to reconsider our original questions: What should a player do? What should happen? But, rather than try to answer these questions now for all games of perfect information, we will restrict ourselves to three very simple cases: (1) A row associated with a strategy for white contains only W's. (2) A column associated with a strategy for black contains only L's. (3) At least one row has no L's, and at least one column has no W's.

In case 1, white's job is easy: he simply picks the right strategy and wins. Case 2 is identical to case 1 except that the roles of the players are reversed. In case 3, by selecting appropriate strategies—a row with no L's and a column with no W's, each player will avoid losing. Since the outcome will be a loss for neither player, it must be a draw.

So, in these three cases at least, it's easy enough to answer our two original questions. But what of a game that does not fall into the pattern of any of these three? Such games certainly exist: the child's game of stone-paper-scissors is one. In this game, two players simultaneously put out their hands in the form of paper, stone, or scissors. Stone beats scissors, scissors beat paper, and paper beats stone. If we denote the scissors, paper, and stone strategies for white and black by S_1, S_2, S_3 and T_1, T_2, T_3, respectively, we obtain the matrix representation of the game shown in Figure 2.

Black's Strategies

		T_1	T_2	T_3
White's	S_1	D	W	L
Strategies	S_2	L	D	W
	S_3	W	L	D

Figure 2

It's easy to see that this game falls into none of the three cases we just considered. Here, advance information of your opponent's strategy would be very valuable; in fact, would guarantee a win.

As it happens, a game of perfect information can never be represented by a matrix like that in Figure 2. Indeed, the stone-paper-scissors game is not one of perfect information, because neither player knows, when he makes his own choice, what his opponent's choice is.

Every two-person, zero-sum, finite game of perfect information is included in one of the three cases given above. This property—that the three cases cover all possibilities—is called "strict determinacy." Ernst Zermelo, in 1912, in what was probably the earliest paper on game theory, proved that games of perfect information are strictly determined. An informal outline of such a proof is included in the appendix to this chapter.

Some Qualifications and Second Thoughts

Games of perfect information are very simple and, seemingly, everything that needs to be said about them is adequately covered by the strict-determinacy theorem. But that isn't so. What is most disturbing is that, despite the theory, game-theorists lose chess games; they lose them when play-

ing with the white pieces and also when playing with the
black ones. And this happens because games that can be
solved only in principle are treated as though they could
be solved in fact.

In their delightful book, *Mathematics and the Imagination*, Edward Kasner and James R. Newman distinguish
mathematics, which uses little words for big ideas, from
other sciences that use big words for little ideas. Now a
strategy is not a profound concept, but it is of tremendous
length. Actually to put a game such as chess in normal
form and compute the outcome is out of the question, even
with the aid of the computers that exist today or the ones
that will be constructed in the foreseeable future. When we
speak of the "solution" that exists for this type of game, we
are ignoring the practical problem of actually finding it.
We should not be surprised when we discover that the
theory has limited practical application.

This attitude is of course not peculiar to game theory.
A person who has mastered long division is not actually required to be able to divide a number with a million digits by
another with ten thousand; it is sufficient if he can describe
how it can be done.

An objection that is often raised is that game theory is all
very well if the players are sensible but is inadequate if the
players are irrational. We will return to this point when we
discuss other games. For games of perfect information, this
objection has no validity whatever, at least not in theory.

If player *A* has a winning strategy against player *B*, and
has the cleverness to discern it and the will to use it, player
B is powerless. Not only is it unnecessary to make any assumption about *B*'s rationality, it is meaningless even to
talk about it. How can you talk of *B*'s rationality when the
outcome of the game will be the same—a loss for *B*—no
matter what *B* does?

Of course, if a player does not have a theoretical win
against his opponent, it makes a great deal of practical difference if his opponent plays well or badly—especially if

the player can foretell the exact nature of the opponent's errors. Against an irrational player (or one who is simply not omniscient), it is important which losing (or drawing) strategy is chosen. (Against an ideal player, of course, it is irrelevant.) It is well known that certain kinds of strategies induce errors on the part of a fallible human opponent. Emanuel Lasker, world champion for many years, felt that psychology plays a very important role in the game of chess. He often adopted a slightly inferior opening, which initially gave him a slight disadvantage, in order to disconcert his opponent. A Russian handbook on championship chess suggests that a player should try to force his opponent into an early commitment, even if by doing so the first player obtains a slightly inferior position. In the children's game of tick-tack-toe, the outcome will always be a draw if there is correct play on both sides, but there is a pragmatic argument for making the first move in the corner: there is only one answer to the corner move that preserves the draw, and that is a move in the center. Any other first move allows at least four adequate replies. So, in a sense, the corner move is strongest, but it is a sense that the game-theorist does not recognize. He does not speak of "slightly unfavorable positions" or "commitments" or "attacks," premature or otherwise. He is incompetent to deal with the game on these terms, and these terms are superfluous to his theory. In short, the game-theorist does not attempt to exploit his opponent's folly.

Since it takes no great insight to recognize the existence of folly in this world, and since the game-theorist purports to be influenced by the world, why this puristic attitude? The answer is simply this: it is much easier to recognize the existence of error than to fashion a general, systematic theory that will allow one to exploit it. So the study of tricks is left to the experts in each particular game; the game-theorist makes the pessimistic, and often imperfect, assumption that his opponent will play flawlessly.

Appendix

Games of Perfect Information Are Strictly Determined

There is a general theorem which states that every two-person, finite, zero-sum game of perfect information is strictly determined. We will outline the proof of a somewhat more restricted theorem. The purpose of including this proof is to clarify further certain of the concepts we used earlier, such as "strategy" and "strictly determined," and to show how these concepts may be applied. To this end, we will keep the argument as simple and informal as possible.

We will be concerned only with games that end in a win for one player and a loss for the other. Other games such as chess which can also end in a draw will be excluded. With some effort the proof might be extended, but the compensations will be too slight to warrant the complications.

What Is to Be Shown

The conclusion we wish to reach is that, under appropriate conditions, the game is *strictly determined*. This means that one of the players has a winning strategy—that is, a strategy which, if chosen, will guarantee him a win, no matter what his opponent does.

For convenience, we will call the players A and B and assume that A moves first. A player has a winning position at X if, once he reaches position X, he has a winning strategy.

The Method of Proof

We will show indirectly that one of the players has a winning strategy. That is, we will show that the assumption that neither player has a winning strategy leads to a contradiction. We will not actually detail a winning strategy; that is a practical impossibility in a game as complicated as chess. We will merely prove the strategy's existence. (It is as though you were told that three families have, among them, a total of seven children; you can be sure that at least one family has more than two children, although you might not know which one.)

We start by assuming that the game is *not* strictly determined. That is, whatever strategy A uses, there is at least one strategy B can select which will lead to a win for B. And whatever strategy B uses, there is a strategy A can pick which will allow A to win. In other words, the initial position is not a win for either player. Two conclusions immediately follow:

1. *Any position to which* A *can move from the initial position will not be a winning position for player* A. If this were not so—if A could go to some position X which was a winning position for himself, he would have had a winning position from the beginning. His strategy would be to go to X and then use the winning strategy that exists at this position.

2. *There is at least one position to which* A *can go from the initial position which is not a winning position for player* B. If this were not true, B would have a win from the start. He need only wait for A to make his first move and then play the winning strategy that exists at his winning position.

Now we are almost done. From 1 and 2 it follows that there is a position immediately after the initial position which is not a win for either player. Call that position Y. In exactly the same way, we can show that there is a position immediately after Y at which neither player has a win.

We can continue in this way indefinitely, and so the game can never end. (Once the game ends, one of the players must have reached a winning position.) This contradicts the assumption that the game is finite, and the theorem is proven.

3

The General, Finite, Two-Person, Zero-Sum Game

In February 1943, General George Churchill Kenney, Commander of the Allied Air Forces in the Southwest Pacific, was faced with a problem. The Japanese were about to reinforce their army in New Guinea and had a choice of two alternative routes. They could sail either north of New Britain, where the weather was rainy, or south of New Britain, where the weather was generally fair. In any case, the journey would take three days. General Kenney had to decide where to concentrate the bulk of his reconnaissance aircraft. The Japanese wanted their ships to have the least possible exposure to enemy bombers, and, of course, General Kenney wanted the reverse. The problem is shown in Figure 3 in normal form; the matrix entries represent the expected number of days of bombing exposure.

Japanese Choice

		Northern	Southern
Allied	Northern	2 days	2 days
Choice	Southern	1 day	3 days

Figure 3

In August 1944, just after the invasion of Europe, the Allies broke out of their beachhead at Cherbourg and were threatening the German Ninth Army. The German commander had this choice: attack or retreat. The Allied commander had these choices: to reinforce the gap, to push his reserves eastward as fast as possible or to hedge—wait twenty-four hours, and then decide whether to reinforce or to push east. For each of these contingencies, the probable outcome is given in Figure 4.

Both these situations* are examples of two-person, zero-sum games.† The critical difference between them and the games of perfect information we discussed earlier is the amount of information available to the players. This new element—imperfect information—complicates matters considerably.

The final outcome of a game of perfect information such as chess depends on the actions of both players; so, at first glance, it would seem reasonable that a player should be concerned with what his opponent will do. But, upon closer

* These two situations were taken from an article by O. G. Haywood, Jr., entitled "Military Decisions and Game Theory." In the first, both the Allied and the Japanese forces chose the rainy route. In the second, the Allies held back for twenty-four hours. The Germans planned to withdraw, but Hitler countermanded the orders and the Germans were almost completely surrounded; only remnants of the German Ninth Army were extricated.

†It is not coincidental that we took both examples from a military context. In most game-like situations there are some cooperative elements—interests common to both players—but in problems of military tactics there often are none.

examination, it turns out that a player can convert a theoretical win or draw into an actual one, at least in principle, without concerning himself with what his opponent does. In chess, if you have a winning (or drawing) strategy, you can win (or draw), no matter what your opponent does; in fact, you can announce your strategy in advance. If your opponent has a winning strategy, on the other hand, you are theoretically helpless; you must resort to trying to trick him to obtain what you cannot against proper play. These are the only kinds of situations that can arise in games of perfect information. For these games, then, if there is sensible play, the outcome is predetermined.

German Actions

		Attack	Retreat
	Reinforce Gap	Allied forces probably repulse attack	Gap holds, but only weak pressure exerted on the retreating Germans
Allied Actions	Move Reserves East	Germans have good chance of sealing gap, cutting off reserves extended east	Allied forces ideally disposed to harass the retreating Germans
	Hold Back Reserves	Gap probably holds, with a chance of encircling the Germans	Reserves arrive one day late, and only moderate pressure is exerted on the Germans

Figure 4

In the absence of perfect information, the situation is more complicated. To see why, consider the stone-paper-scissors game we referred to earlier. This is *not* a game of perfect information, because the players must choose their strategies simultaneously, neither knowing what the other is going to do. In this game, if you know what your opponent will do, it is a simple matter to win. He has only one of three choices: stone, paper, or scissors; and your respective winning answer is paper, scissors, or stone. Whichever one of these strategies you adopt is best kept to yourself.

The conclusion we just reached—that no strategy in the game of stone-paper-scissors is any good unless it is kept secret—would seem to make it impossible to choose a strategy intelligently. To try to construct a theory for such a game would appear to be a waste of time. Von Neumann and Morgenstern put the problem this way: "Let us imagine that there exists a complete theory of the zero-sum two-person game which tells a player what to do and which is absolutely convincing. If the players knew such a theory then each player would have to assume that his strategy has been 'found out' by his opponent. The opponent knows the theory, and he knows that a player would be unwise not to follow it. Thus the hypothesis of the existence of a satisfactory theory legitimatizes our investigation of the situation when a player's strategy is 'found out' by his opponent."

The paradox is this: if we are successful in constructing a theory of stone-paper-scissors that indicates which of the three strategies is best, an intelligent opponent with access to all the information available to us can use the same logic to deduce our strategy. He can then second-guess us and win. So it would be fatal to use the "preferred" strategy suggested by the theory. Before we try to resolve this paradox, let's first go back and consider a few simple examples.

A Political Example

It is an election year and the two major political parties are in the process of writing their platforms. There is a dispute between state X and state Y concerning certain water rights and each party must decide whether it will favor X or favor Y or evade the issue. The parties, after holding their conventions privately, will announce their decisions simultaneously.

Citizens outside the two states are indifferent to the issue. In X and Y, the voting behavior of the electorate can be predicted from past experience. The regulars will support their party in any case. Others will vote for the party supporting their state or, if both parties take the same position on the issue, will simply abstain. The leaders of both parties calculate what will happen in each circumstance and come up with the matrix shown in Figure 5. The entries in the matrix are the percentage of votes party A will get if each party follows the indicated strategy. If A favors X, and B

B's Platform

		Favor X	Favor Y	Dodge Issue
	Favor X	45%	50%	40%
A's	Favor Y	60%	55%	50%
Platform	Dodge Issue	45%	55%	40%

Figure 5

dodges the issue, A will get 40 percent of the vote.

This is the simplest example of this type of game. Though both parties have a hand in determining how the electorate will vote, there is no point in one party trying to anticipate what the other will do. Whatever A does, B does best to dodge the issue; whatever B does, A does best to support Y. The predictable outcome is an even split. If, for some reason, one of the parties deviated from the indicated strategy, this should have no effect on the other party's actions. A slightly more complicated situation arises if the percentages are changed a little as shown in Figure 6.

B's Platform

	Favor X	Favor Y	Dodge Issue
Favor X	45%	10%	40%
A's Favor Y	60%	55%	50%
Platform Dodge Issue	45%	10%	40%

Figure 6

B's decision is now a bit harder. If he thinks A will favor Y, he should dodge the issue; otherwise, he should favor Y. But the answer to the problem is in fact not far off. A's decision is clear-cut and easy for B to read: favor Y. Unless A is a fool, B should realize that the chance of getting 90 percent of the vote is very slim—indeed, not a real possibility—and he would do best to dodge the issue.

This is the same type of situation that General Kenney had to contend with. On the face of it, the northern and southern routes both seemed plausible strategies. But the rainy northern route was obviously more favorable for the Japanese, which meant that the northern route was the only reasonable strategy for the Allies.

In Figure 7, neither player has an obviously superior strategy. In this case, both players must think a little. Each player's decision hangs on what he expects the other will do. If B dodges the issue, A should too. If not, A should favor Y. On the other hand, if A favors Y, B should favor X. Otherwise, B should favor Y.

Once again, the underlying structure is not hard to analyze. While at first B may not be clear what he should do, it's obvious what he should *not* do: he should not dodge the issue, since whatever A does, B always does better if he favors X over dodging the issue. Once this is established, it immediately follows that A should favor Y and, finally, that B should favor X. A will presumably wind up with 45 percent of the vote.

These two strategies—A's favoring Y, and B's favoring X—are important enough to be given a name: *equilibrium strategies*. The outcome resulting from the use of these two strategies—the 45 percent vote for A—is called an *equilibrium point*.

B's Platform

		Favor X	Favor Y	Dodge Issue
	Favor X	35%	10%	60%
A's	Favor Y	45%	55%	50%
Platform	Dodge Issue	40%	10%	65%

Figure 7

What are equilibrium strategies and what are equilibrium points? Two strategies are said to be in equilibrium (they come in pairs, one for each player) if neither player gains by changing his strategy unilaterally. The outcome (sometimes called the payoff) corresponding to this pair of strategies is defined as the equilibrium point. As the name suggests, equilibrium points are very stable. In two-person, zero-sum games, at any rate, once the players settle on an equilibrium point, they have no reason for leaving it. If, in our last example, A knew in advance that B would favor X, he would still favor Y; and, similarly, B would not change his strategy if he knew A would favor Y. There may be more than one equilibrium point, but if there is, they will all have the same payoff.

This is true in two-person, zero-sum games, that is. One also speaks of equilibrium strategies and equilibrium points in n-person and non-zero-sum games, but the arguments favoring them are considerably less appealing. In a two-person, non-zero-sum game, for example, equilibrium points need not have the same payoff; one equilibrium point may be more attractive to *both* players than another. We will discuss this in the "prisoner's dilemma."

When two-person, zero-sum games have equilibrium points, it is generally agreed that their common payoff should be the final outcome. But there are also games with no equilibrium points, and it is these that cause the difficulty. Edgar Allan Poe discussed such a game—an extremely simple one —in *The Purloined Letter,* and it is interesting to compare his approach with von Neumann's, which we will discuss later. Here is Poe:

"I knew one about eight years of age, whose success at guessing in the game of 'even and odd' attracted universal admiration. This game is simple, and is played with marbles. One player holds in his hand a number of these toys, and demands of another whether that number is even or odd. If the guess is right, the guesser wins one; if wrong, he loses one. The boy to whom I allude won all the marbles of the school. Of course he had some principle of guessing; and this lay in mere observation and admeasurement of the astuteness of his opponents. For example, an arrant simpleton is his opponent, and, holding up his closed hand, asks, 'are they even or odd?' Our schoolboy replies, 'odd,' and loses; but upon the second trial he wins, for he then says to himself, 'the simpleton had them even upon the first trial, and his amount of cunning is just sufficient to make him have them odd upon the second; I will therefore guess odd;'—he guesses odd, and wins. Now, with a simpleton a degree above the first, he would have reasoned thus: 'This fellow finds that in the first instance I guessed odd, and, in the second, he will propose to himself, upon the first impulse, a simple variation from even to odd, as did the first simpleton; but then a second thought will suggest that this is too simple a variation, and finally he will decide upon putting it even as before. I will therefore guess even;'—he guesses even, and wins. Now this mode of reasoning in the schoolboy, whom his fellows termed 'lucky,'—what, in its last analysis, is it?"

"It is merely," I said, "an identification of the reasoner's intellect with that of his opponent."

"It is," said Dupin; "and, upon inquiring of the boy by what means he effected the *thorough* identification in which his success consisted, I received answer as follows: 'When I wish to find out how wise, or how stupid, or how good, or how wicked is any one, or what are his thoughts at the moment, I fashion the expression of my face, as accurately as possible, in accordance with the expression of his, and then wait to see what thoughts or sentiments arise in my mind or heart, as if to match or correspond with the expression.' "

. . .

"As poet *and* mathematician, he would reason well; as mere mathematician, he could not have reasoned at all, and thus would have been at the mercy of the Prefect." *

It will be easier to appreciate von Neumann's approach to this problem if we look a little more at the problem first. Put yourself in the shoes of one of the schoolboys who was pitted against Poe's remarkable wonder child.

Your situation seems hopeless; every idea that occurs to you will occur to the wonder child too. You can try to fool him by fixing your face into an "even configuration" and then playing odd. But how do you know that what you think of as an "even" configuration is not really the expression of someone playing odd and trying to look as though he were playing even? Perhaps you ought to look even, think even, and then (slyly) *choose* even. But once again, if you are clever enough to figure out this tortuous plan, won't he be clever enough to see through it? This kind of reasoning can be extended to any number of steps and the effort would be as futile as it is exhausting.

Let's look at the game in normal form as shown in Figure 8. The numbers in the matrix indicate the number of marbles that you are paid by the wonder child (or that you pay, if the number is negative).

Now, instead of trying to figure out just how much insight the wonder child has, suppose you assume the worst—he is so clever that he can anticipate your thinking on every count. In effect, you must announce your strategy in advance, and the wonder child can use the information as he wishes. In this case, it seems to make little difference what

* Quoted from Edgar Allan Poe, *Poe's Tales* (New York, 1845).

you do. Whether you choose even or odd, the result will be the same: the loss of a marble. Thus, if the wonder child has perfect insight, it appears that you may as well resign yourself to the loss of a marble. You certainly can't do any worse than this—but is there a chance of doing better?

The fact is that you *can* do better, despite your opponent's cleverness. And, ironically, the means of doing so is inadvertently suggested by our second quotation from Poe: " . . . *The way to do better is not to reason at all.*" To see why, let's go back a bit.

Poe's Student

		Even	Odd
You	Even	−1	+1
	Odd	+1	−1

Figure 8

We said earlier that, as chooser, you have only these alternatives: to pick an even number of marbles or an odd number of marbles. In one sense this is true; ultimately, you have to make one of these two choices. But in another sense this is *not* true. There are many *ways* in which you can make this choice. And although it may seem that *what* you decide is all that matters and *how* you decide is irrelevant, how you decide is actually critical. Of course, you can always pick one of the *pure* strategies: odd or even. But you can also use a chance device such as a die or roulette wheel to make the decision for you. Specifically, you might throw a die and if a six appears choose even and otherwise choose odd. A strategy that requires the selection of a pure strategy by means of a random device is called a *mixed* strategy.

From the second point of view, you have not two but an infinite number of mixed strategies: you can choose an odd number of marbles with probability p and an even number of marbles with probability 1-p (p being a number between zero and one). Roughly speaking, p is the fraction of times an odd number would be chosen if the game were played an infinite number of times. The pure strategies—"always play even," "always play odd"—are the extreme cases, where p is zero and 1, respectively.

Returning to *The Purloined Letter*, let us suppose you play even half the time; you might flip a coin and play odd whenever it comes up heads. Suppose the wonder child guesses that p is one half, or you tell him. There is nothing more he can learn from you—you simply don't know any more. Also, unless he has powers undreamed of even by Poe, he can't possibly predict the outcome of a random toss of a coin. If you pick this mixed strategy, the outcome will be the same no matter what your opponent does; that is, each of you will win, on the average, half the time.

To summarize: you have taken a game in which you were seemingly at the mercy of a clever opponent, and transformed it into one in which your opponent had essentially no control over the outcome. But if you can do as well as this when you give away your strategy, wouldn't you do better if you kept it to yourself? No, you can't do any better, and it's clear why: your opponent can do the same thing you have done. By using a random device, he, too, can ensure an even chance of winning. So *each* player has it within his power to see that no advantage is gained by the other.

It is conceivable that a player may do better by not "randomizing." In the game described by Poe, for example, if a player has a talent for anticipating his opponent's choices, he might try to exploit that talent. If your opponent doesn't know about randomizing (or, mistakenly, feels he is cleverer than you), it may be possible to exploit his ignorance. But this is a doubtful advantage at best: it can always be neutralized by a knowledgeable opponent.

It should also be noted that in the odd-even game, once a player resorts to randomizing, not only can he not lose (on average), no matter how well his opponent plays; he will not gain, *however badly his opponent plays.* The more capable your opponent, then, the more attractive the randomizing procedure.

The reasoning on which we based our analysis of the odd-even game is applicable in much more complicated situations as well.

A Military Application

General X intends to attack one or both of two enemy positions, A and B; and General Y must defend them. General X has five divisions at his disposal; General Y has three. Each general must divide his entire force into two parts and commit them to these two positions, without knowing what the enemy will do. The outcome is determined once the strategies are chosen. The general who commits the most divisions to any one position will win there; if both generals send the same number of divisions, each will get credit for half a victory. Assuming that each general wishes to maximize his expected number of victories (that is, the average), what should he do? What is the expected outcome?

The entries in the matrix in Figure 9 represent the average number of victories won by General Y. If General Y sends two divisions to A, for example, and General X does the same, General Y will gain half a victory at A and no victory at B, where he is outnumbered three to one.

Let's start by looking at the problem from General Y's point of view. Assume he's a pessimist and he believes General X is capable of second-guessing him. It is clear that whatever strategy he picks, he will have no victories; General X will assign one more division to A than General Y did and have one extra division at B as well.

Now suppose General Y decides to play a mixed strategy; that is, he makes his decision on the basis of a chance device.

We will still assume that General X can guess the nature of the chance device (the probabilities of selecting each of the strategies), but not the actual outcome of the toss of the coin or spin of the wheel. Specifically, suppose General Y assigns all his divisions to A one-third of the time, all his divisions to B one-third of the time, and makes each of the two possible two/one splits one-sixth of the time.*

X's Strategies

Divisions Sent to A

			0	1	2	3	4	5
	Divisions	0	1/2	0	1/2	1	1	1
Y's	Sent	1	1	1/2	0	1/2	1	1
Strategies	to	2	1	1	1/2	0	1/2	1
	A	3	1	1	1	1/2	0	1/2

Figure 9

General X, aware of General Y's strategy, need examine only what will happen in each of his own six strategies and then select the one that gives him the greatest advantage. It is a matter of simple calculation to determine that General X will obtain an average of 1 5/12 victories (and, correspondingly, General Y will obtain an average of 7/12 victories) if he does not put all his divisions in one location, and he will obtain an average of 1 1/6 victories otherwise.

What can we conclude from all this? Simply that General Y can reasonably expect to win an average of 7/12 victories, no matter how clever General X is. In fact, he can win the 7/12 victories even if he tells General X his strategy in advance. But should he settle for this? To answer this ques-

---

* We are primarily concerned with describing the meaning and significance of the von Neumann solution. The technique of computing the strategies and value of the game will be found in more technical presentations. Consequently, we will pull solutions out of a hat without apology whenever appropriate.

tion, let us look at the problem from General X's point of view.

It is clear from the start that if General X commits himself to a pure strategy and General Y figures it out, they will each get one victory. When five divisions are distributed between two locations, there must be two or less divisions at one of the locations. If General Y sends all three of his divisions to the location where General X sent his smaller force, he will gain precisely one victory (and there is no way he can do any better).

But now suppose General X sends one, two, three, four divisions to A, with probabilities 1/3, 1/6, 1/6, 1/3, respectively, and the balance of his forces to B. (A little thought should make it obvious that it never pays to send all five divisions to one location.) It makes no difference what General Y does. The result will be the same: General Y will obtain an average of 7/12 victories.

In short, the situation is this. Either General X or General Y can, without the aid of, and without errors from, his opponent, ensure an outcome at least as favorable to himself as the outcome mentioned above: 7/12 of a victory (on average) for General Y, 1 5/12 victories for General X. The answer to our earlier question is that neither general can hope to do better than this against an informed opponent.

A Marketing Example

Two firms are about to market competitive products. Firm C has enough money in its advertising budget to buy two blocks of television time, an hour long each, while firm D has enough money for three such one-hour blocks. The day is divided into three basic periods: morning, afternoon, and evening, which we will indicate by M, A, and E, respectively. Purchases of time must be made in advance and are kept confidential. Fifty percent of the television audience watch in the evening; 30 percent watch in the afternoon; and the remaining 20 percent watch in the morning. (For the sake of simplicity, we assume that no person watches during more than one period.)

C's Strategies

		EE	EA	EM	AA	AM	MM
	EEE	75	60	65	60	50	65
	EEA	65	75	80	60	65	80
	EEM	60	70	75	70	60	65
	EAA	40	65	55	75	80	80
D's	EAM	50	60	65	70	75	80
Strategies	EMM	35	45	60	70	70	75
	AAA	40	40	30	65	55	55
	AAM	50	50	40	60	65	55
	AMM	50	35	50	45	60	65
	MMM	35	20	35	45	45	60

Figure 10

If a firm buys more time during any period than its competitor, it captures the *entire* audience during that period. If both firms buy the same number of hours during any one period—and this is so if neither firm buys time at all—each gets half the audience. Each member of the audience buys the product of just one of the firms. How should the firms allocate their time? What part of the market should each firm expect to get?

The game is expressed in normal form in Figure 10. The strategies for firm *C* (there are six) are indicated by two letters. *MA* means one hour of advertising in the morning and another hour in the afternoon. Similarly, the ten strategies for firm *D* are expressed by three letters. The entries in the matrix represent the percentage of the market that firm *D* captures; firm *C* gets the rest.

To illustrate how the percentage entries were obtained,

assume firm D chooses *EMM*—one evening and two morning hours—and firm C chooses *EM*—one evening and one morning hour. Since firm D bought two morning hours to firm C's one, firm D will capture all of the morning market of 20 percent. Since each firm has one evening hour and no afternoon hour these markets of 50 percent and 30 percent, respectively, will be divided equally between the two firms. Thus, firm D will get 60 percent of the total market if the indicated strategies are used.

The Solution to the Marketing Example

It often happens that some strategies are clearly bad and can be rejected immediately; this is the case here. Firm D would never play strategy *AAA*, since it would do better to play *EEM*, no matter what firm C does. *EEM* is said to dominate *AAA*.

A strategy *dominates* another when, whatever the action taken by the opponent or opponents, the first strategy leads to an outcome at least as favorable as the second and, in at least one strategy of the opponent or opponents, actually leads to something better. There is never an advantage to playing a strategy that is dominated.

In the present case, *EEM* dominates *AMM* and *MMM* as well as *AAA*, and *EAM* dominates *EMM* and *AAM*. This means that firm D's last five strategies can be effectively discarded. Once this is done, *AM* dominates *MM*.* (If firm D plays *AAM*, firm C would do better with *MM* than with *AM*; but we have already decided that firm D won't play *AAM*.) The game, then, is reduced to one in which each player has five strategies.

One solution to this problem—and, again, we won't say how we got it—is that firm D should play each of the strat-

---

* Since the entries in the matrix represent the percentages that firm D gets, firm C wants these to be low.

egies *EEE*, *EEA*, and *EAM* 1/3 of the time, and firm *C* should play *EE* 6/15 of the time, *AA* 5/15 of the time, and *AM* 4/15 of the time. If firm *D* uses its recommended strategies, it can be sure of winning on average at least 63 1/3 percent of the audience; and if firm *C* uses its recommended strategy, firm *D* won't win any more than this.

Simplified Poker

A and *C* each puts $5 on the table and then toss a coin which has *1* on one side and *2* on the other. Neither player knows the outcome of the other's toss.

A plays first. He may either pass or bet an additional $3. If he passes, the numbers tossed by the two players are compared. The larger number takes the $10 on the table; if both numbers are the same, each gets his $5 back.

If *A* bets $3, *C* can either see or fold. If *C* folds, *A* wins the $10 on the table, irrespective of the numbers tossed. If *C* sees, he adds $3 to the $13 already on the table. Again the numbers are compared; the larger number takes the $16, and if the numbers are equal, each gets his money back. What are the best strategies, and what should happen?

Each player has four strategies. *A* can always pass (indicated as *PP*), pass with *1* and bet with *2* (*PB*), pass with *2* and bet with *1* (*BP*), or always bet (*BB*). *C* can always fold (*FF*), always see (*SS*), see with *1* and fold with *2* (*SF*), and fold with *1* and see with *2* (*FS*).

To illustrate how the entries in the payoff matrix in Figure 11 were obtained, suppose *A* plays (*BB*) and *C* plays (*SF*). Half the time, *C* will get 2 and fold (*A* always bets), and *A* will win $5. A fourth of the time, *C* will get *1* and *A* will get *1* and nothing will happen. And a fourth of the time, *C* will get *1* and *A* will get *2*; *A* will bet (as always), *C* will see, and *A* will win $8. *A*'s average gain, if these two strategies are adopted, is $4.50.

We notice first that *BB* dominates *BP* and *PP*, while *FS*

C's Strategies

		FF	SS	SF	FS
	PP	0	0	0	0
A's	BB	5	0	4 1/2	1/2
Strategies	PB	5/4	3/4	2	0
	BP	3 3/4	−3/4	5/2	1/2

Figure 11

dominates *SF* and *FF*. The formal model reflects what is probably intuitively clear: whenever *A* gets 2 on the toss, he should bet; and whenever *C* gets 2 on the toss, he should see. The strategy recommended for *A* is to play *BB* 3/5 of the time and *PB* 2/5 of the time. This will assure *A* an average win of 30 cents a play. If *C* plays *FS* 3/5 of the time and *SS* 2/5 of the time, *C* will have a loss of no more than 30 cents, on average, per play.

An Old Problem with a New Variation

This problem is similar to the one we discussed earlier, in which five criminals are meeting, while a policeman is waiting outside to apprehend the leader. But now the leader knows about the policeman. (The leader does not tell his comrades, however, because it was his own error that attracted the policeman.) The members of the gang leave in a random order, just as they did before, but the leader selects his time of departure. What are the best strategies for the policeman and for the gang leader, and, assuming *they use them,* what is the probability of a successful getaway?

The pure strategy that we recommend for the policeman is no longer adequate. When the policeman sees a gang member leave who is taller than anyone who left earlier he should not simply decide whether to follow him or not but he should, rather, follow him with a certain probability. And this probability rises as more people leave. Specifically, the policeman should follow the first person who leaves 12/37 of the time and the second, third, fourth, and fifth person to leave (assuming in each case the person departing is taller than his predecessors) 12/25, 12/19, 4/5, and all of the time, respectively. It turns out that if the policeman follows this strategy the probability of his following the leader is 12/37 and it makes no difference when the leader actually leaves.

The leader should also randomize. He should leave first 12/37 of the time. If he doesn't leave first he should leave second 12/25 of the time. If he is still around he should leave third, fourth and fifth with a probability of 12/19, 4/5, and 1, respectively.

This game and the policeman vs. gangsters game we discussed earlier seem very similar, but actually they are quite different. In the one-person game, the policeman's presence was unknown, and nature—which determined the order of the gangsters' departure—was neutral. But the policeman in the two-person game is playing against an adversary who wants to escape, and not against indifferent nature, and he must act accordingly. He cannot let the first few gang members pass, as he did in the one-person game, for if he did, the gang leader would be certain to escape by leaving first. Also, the policeman cannot expect to do as well against a hostile opponent as he would against nature. He doesn't: his chance of catching the gang leader drops from 13/30 to 12/37.

It is interesting to see what happens when the number of gangsters is increased from five to a much larger number. We mentioned earlier that in the one-person game a clever policeman will pick out the gang leader about 40 percent of the time, no matter how many criminals there are. In the two-person game, he does considerably worse. As the number of gangsters increases, his chance of picking the leader approaches zero. More precisely, if there are n gangsters, the chance of catching the leader is about $1/(\log_e n)$.

This example is not as artificial as it might seem. It emerged, in a modified form, from the construction of models in connection with a hypothetical nuclear-test-ban treaty. The models were to determine how efficient certain inspection procedures would be in verifying that the treaty was being observed. The "criminals" were natural disturbances such as earthquakes; the "gang leader" was an illegal test carried out by a violator of the treaty; and detecting the leader was detecting the site of the questionable disturbance.

The Minimax Theorem

Up to now we have considered specific games only. In each of them we recommended certain strategies for the players and indicated what the outcome should be. All these examples were special instances of one of the most important and fundamental theorems of game theory: von Neumann's minimax theorem.

The minimax theorem states that one can assign to every finite, two-person, zero-sum game a value V: the average amount that player I can expect to win from player II if both players act sensibly. Von Neumann felt that this predicted outcome is plausible, for three reasons:

1. There is a strategy for player I that will protect this return; against this strategy, nothing that player II can do will prevent player I from getting an average win of V. Therefore, *player I will not settle for anything less than* V.

2. There is a strategy for player II that will guarantee that he will lose no more than an average value of V; that is, *player I can be prevented from getting any more than* V.

3. By assumption, the game is zero-sum. What player I gains, player II must lose. Since player II wishes to mini-

mize his losses, *player* II *is motivated to limit player* I's *average return to* V.

The last assumption, which is easily overlooked, is crucial. In non-zero-sum games, where it does not hold, one should not conclude that just because player *II* has the power to limit player *I*'s gains, he will necessarily do so. But here, where hurting player *I* is, for *II*, equivalent to furthering his own ends, the assumption is compelling.

The concept of mixed strategies and the minimax theorem considerably simplify the study of these games. To appreciate how much, imagine what the theory would be like in their absence. Except for the simplest types of games—games in which there are equilibrium points—there is anarchy. It is impossible to choose a sensible course of action or predict what will happen; the outcome is intimately related to *both* players' behavior and each is at the mercy of the other's caprice. One can only try to define rational play along the lines suggested by Poe, and such attempts are virtually worthless.

With the addition of the minimax theorem, the picture is radically different. In effect, one can treat all two-person, zero-sum games as though they had equilibrium points. The game has a clear value, and either player can enforce this value by selecting the appropriate strategy. The only difference between games with actual equilibrium points and those without them is that in one case you can use a pure strategy and obtain the value of the game with certainty, while in the other case you must use a mixed strategy and you obtain the value of the game on average.

The virtue of the minimax strategy is security. Without it, you must resort to the double- and triple-cross world of Poe's precocious student. With it, you can obtain your full value, and you have the assurance that you couldn't do better—at least, not against good play.

Some Second Thoughts

Most knowledgeable game-theorists, when asked to select the most important single contribution to game theory, would probably choose the minimax theorem. The arguments in favor of minimax strategies are very persuasive; still, they can be overstated. And even a sophisticated writer, in a careless moment, can fall into a trap. In his book, *Fights, Games and Debates*, Anatol Rapoport discusses the game shown in Figure 12. The numbers in the payoff matrix indicate what player *II* pays player *I*. The units are unimportant as long as we assume that each player would like to receive as much as he can.

Player *II*

	A	B
a	−1	5
b	3	−5

Player *I*

Figure 12

Rapoport starts by making the routine calculations. He computes player *I*'s minimax strategy (play strategy *a* 4/7 of the time and strategy *b* 3/7 of the time), player *II*'s minimax strategy (play strategy *A* 5/7 of the time and strategy *B* 2/7 of the time), and the value of the game (an average win of 5/7 for player *I*). He then goes on to say: "To try to evade by using a different strategy mixture can only be of disadvantage to the evading side. There is no point at all in trying to delude the enemy on *that* score. [Rapoport's italics] Such attempts can only backfire." But Rapoport's conclusion that any deviation from the minimax "can only be of disadvantage" leads to a curious paradox. To see why, observe two basic facts.

Note first that if *either* player selects his minimax strategy, the outcome will be the same: an average gain of 5/7 for player *I*. There is only one way to obtain another outcome: *both* players must deviate from their minimax strategies.

The second point is that the game is zero-sum. A player cannot win unless he wins from his opponent; a player cannot lose unless there is a corresponding gain for his opponent.

Putting these two facts together, we have the contradiction. If a player deviates from the minimax, it cannot possibly be "of disadvantage" unless his opponent deviates as well. And, by the same argument, his opponent's deviation must also be of disadvantage. But the game is zero-sum: both players cannot lose simultaneously. Obviously, something is wrong.

Here is the answer. There is a strategy for player *I* that guarantees him 5/7, and he can be stopped from getting any more; moreover, player *II* is motivated to stop him from getting more. If player *I* chooses another strategy, he's gambling. If player *II* also gambles, there is no telling what will happen. The minimax strategies are attractive in that they offer security; but the appeal of security is a matter of personal taste.

As a consequence of the minimax theorem, the general, zero-sum, two-person game has a good theoretical foundation. But, like the game of perfect information, it rarely exists in practice. The difficulty is the requirement that the game be zero-sum.

The essential assumption upon which the theory is based is the opposition of the two players' interests. To the extent that the assumption is not valid, the theory is irrelevant and misleading. Often the assumption seems to be satisfied but in reality is not. In a price war, for example, it may be to the advantage of *both* parties that prices be maintained.

In a friendly game of poker, *both* players may prefer that neither suffers an excessive loss. And when bargaining, buyer and seller may have divergent interests with respect to the price, but *both* may prefer to reach some agreement.

The theory has some applications, however. Various forms of poker have been analyzed with the aid of the minimax theory, and von Neumann and Morgenstern, in their classic *Theory of Games and Economic Behavior*, include a chapter entitled "Poker and Bluffing." Others have also tried their hand at poker, but the models are invariably simplified; as a practical matter, the game is too complex to analyze completely. Certain aspects of bridge, such as signaling, have also been analyzed, and mixed strategies have been applied to bridge play; but no overall model, encompassing both bidding and play, has been devised.

Military tactics, an area in which the parties are involved in almost pure conflict, is another source of two-person, zero-sum games. Evasion and pursuit games and fighter-bomber duels are two applications. In addition, there is a general class of games called Colonel Blotto games, in which the players are required to allocate certain resources at their disposal. We have seen one example of this in a military context—Generals X and Y allocating their divisions—and another in a marketing context. The same problem arises when allocating salesmen to various territories or policemen to high-crime areas. An interesting application of Colonel Blotto games arose in the process of creating a model for inspection in a possible disarmament agreement. One country has factories that produce certain types of weapons —some more potent than others; some easier to hide than others, and so on. The inspecting country has to decide how to allocate its inspectors to verify that the inspected country is not exceeding its quota of weapons.

An interesting application of minimax strategies was observed by Sidney Moglewer. In the process of selecting crops, one can regard the farmer as one player and the "hypothetical combination of *all* the forces that determine market prices for agricultural products" as the other player. Moglewer points out that though it is difficult to justify the farmer's implicit assumption that the universe is concerned with him as an individual, the farmer acts as though this were the case.

Some Experimental Studies

Although we speak of game *theory*, it is important to look at games in practice. In a speech to future jurists at the Albany Law School, Justice Benjamin Cardozo remarked, "You will study the life of mankind, for this is the life you must order, and to order with wisdom, know." The comment applies to game theory as well. Game theory has its roots in human behavior; if the theory is not related to human behavior in some way, it will be sterile and meaningless except as pure mathematics. Aside from other considerations, experiments in game theory are interesting in themselves, perhaps because people enjoy reading about what other people do. This is sufficient reason for discussing them.

Richard Brayer had subjects repeatedly play the game shown in Figure 13. The entries in the payoff matrix indicate what the subject received from the experimenter. The players either were told they were playing against an experienced player or they were told that their opponent would

Experimenter

		A	B	C
	a	11	−7	8
Subject	*b*	1	1	2
	c	−10	−7	21

Figure 13

play randomly, but in fact they invariably played against the experimenter.

It takes only a moment for a subject to deduce that he should play strategy b if he feels his opponent is intelligent. From the point of view of the experimenter, strategy C is dominated by B, so the subject can assume that the experimenter will never play C. Having decided this, the subject should never play c, since he always does better with strategy b. Eliminating c and C, the experimenter should play B, and the subject should play b. (B, b) are equilibrium strategies, and the payoff 1 is an equilibrium point.

What actually happened? For one thing, the subjects ignored what they were told about their opponent and responded only to how he played. Whether they did not see the relevance of what they were told about their opponent, or did not know how to apply it, or whether they were reacting with the natural skepticism subjects have toward anything they are told by an experimenter is not clear. The subjects did play b if the experimenter played B, but not otherwise. Post-experimental interviews confirmed what the play indicated: the players couldn't anticipate the experimenter's choice of strategy B. In fact, *more than half the subjects felt that the experimenter was stupid for playing B and settling for a loss of 1.* When the experimenter picked his strategy at random, the subjects generally responded by playing strategy a: the one that gave them the highest average return.

The same pattern was observed by other experimenters. Several concluded that most subjects are simply unable to put themselves in the shoes of their opponents. Subjects tend to prefer strategies that yield an apparently high average return (against an opponent playing at random) to equilibrium strategies. When players are punished for deviating from the equilibrium strategy, they change their behavior, but not otherwise.

In games without equilibrium points, the players have

even less insight. In an interesting review of experimental games, Anatol Rapoport and Carol Orwant suggest that equilibrium points, when they exist, will be found eventually, if not immediately, by the players. The speed with which they are discerned varies with the experience and sophistication of the players and the complexity of the game. When there are no equilibrium points, when what is called for is a mixed strategy, the situation is much worse. Not only are all but the most sophisticated players incapable of making the necessary calculations, but almost all players do not even see the need for them.

Is the failure to play an equilibrium strategy, where one exists, necessarily proof of a player's ignorance? Perhaps not on the face of it. It might be argued that the case for the equilibrium strategy is based to some extent on the assumption that one's opponent will play reasonably well; if this is not true, a player often does better by not playing the equilibrium strategy. When the experimenter played the equilibrium strategy, the subjects did too. When the experimenter played randomly and "irrationally," however, the subjects persisted in playing non-equilibrium strategies.

All this sounds plausible but, as we mentioned earlier, it isn't what happened. By their own admission after the experiments, the players indicated that they had no insight into what was happening. They "learned" how to react effectively to the specific behavior of the experimenter (all the while believing the equilibrium strategy he chose was foolish), but the subjects' ideas of what the game was about were as muddy at the end as they were at the start.

Apropos of "learning" how to play games, there is a theorem that is interesting enough to warrant a digression. Suppose that two people are playing a game repeatedly but do not have the skill to compute the minimax strategies. They both play randomly the first time, and subsequently they learn from their experience, in the following sense. They each assume that their opponent is playing a mixed strategy with probabilities proportional to the actual frequencies with which he has chosen his strategies in the past. On the basis of this assumption, they play the pure strategy that will maximize their average return. Julia Robinson proved that, under these conditions, both strategies will approach the minimax strategies.

Of what practical significance are these experiments? Does it make any difference to a player to know that players often act irrationally? It didn't matter in the games of perfect information. Does it matter now?

It is always possible to get the value of the game by playing minimax. That is so, whether one's opponent is rational or not. The reason a player is satisfied to get the value of the game is that he knows a clever opponent can stop him from getting any more. But if a player has reason to believe that his opponent will play badly, why not try to do better?

By using the minimax strategy, a player will avoid doing anything foolish, such as playing a dominated strategy. In our simplified poker game, he would always bet if he were dealt a 2; in the marketing game, he would never select three morning hours to advertise; and, in the military game, he would never send five divisions to one location. It should be emphasized, however, that the minimax strategy is essentially defensive, and when you use it, you often eliminate any chance of doing better. In Poe's odd-even game, in the policeman vs. the gangsters game, and in the game discussed by Rapoport, if a player adopts the minimax strategy, he obtains the value of the game; no less, but *no more*.

Suppose that, in the odd-even game, your opponent tends to choose even more often than he should. How can you exploit this if the game is played only once and you don't know how he plays? Even if you know the game is so complex that he is not likely to analyze it correctly, how can you anticipate in what direction he will err? In general, you may suspect that your opponent will pick the strategy that seems to yield the largest average payoff; but if you act on your suspicion, you become vulnerable to an opponent who is one step ahead of you and who is out to get rich by exploiting the exploiter. And if the game is played repeatedly and you manage to do better than you should, your opponent will eventually learn to protect himself, so your advantage is unstable.

The weakest part of our theory is undoubtedly the assumption that a player should always act so as to maximize his average payoff. The justification for the assumption is that, in the long run, not only the average return but the actual return will be maximized. But if a game is played only once, are long-run considerations relevant? John Maynard Keynes made the point in *Monetary Reform:* ". . . Now 'in the long run' this is probably true. . . . But this *long run* [Keynes's italics] is a misleading guide to current affairs. *In the long run* [Keynes's italics] we are all dead." Often a strategy which maximizes the average return is not desirable, much less compelling. Is it irrational to opt for a sure million dollars rather than take an even chance of getting $10 million? This is not just an incidental objection; it goes to the heart of the matter. Let's take a closer look at this problem. In the game in Figure 14, the payoffs are in dollars.

Player *II*

		A	B
Player *I*	a	1 million	1 million
	b	10 million	Nothing
	c	Nothing	10 million

Figure 14

The minimax strategy for player *I* is to play *b* and *c*, with a probability of one-half. Using this strategy, he will get $10 million half the time, and half the time he will get nothing. But player *I* may prefer to have a sure million; in fact, rather than gamble, both players may prefer that player *I* win a sure million. This is one reason why cases are often settled out of court.

Things are not always what they seem, however. We started with what appeared to be a zero-sum game (it was a zero-sum game in dollars), but in fact the players had certain common interests. Yet the critical assumption in zero-sum games is that the *players have diametrically opposed interests.* And this must be so not only for the entries in the payoff matrix, which indicate the outcome when each player uses a pure strategy, but also for the various probability distributions that may be established when the players use mixed strategies.

This objection is almost fatal. It can be overcome only by introducing an entirely new concept, the concept of utility. (The word is old, but the concept is new.) The utility concept puts the theory on a firm foundation once again and in fact is one of von Neumann and Morgenstern's most significant contributions to game theory. We will discuss it next.

4
Utility Theory

> It is true that only one out of a hundred wins,
> but what is that to me?
> —Fyodor Dostoevsky, *The Gambler*

> "Would you tell me, please, which way I
> ought to go from here?"
> "That depends a good deal on where you
> want to get to," said the cat.
> "I don't much care—" said Alice.
> "Then it doesn't matter which way you
> go," said the cat.
> —Lewis Carroll, *Alice in Wonderland*

In the last chapter we were primarily concerned with the minimax theorem—why it is needed, what it means, and why it is important. For the sake of clarity, we discussed the central ideas and avoided mentioning the difficulties. At the end of the chapter, however, it was suggested that all was not as it should be. Part of the foundation, and an absoutely essential part, is missing, and the theory is in fact firmly rooted in mid-air. One of von Neumann and Morgenstern's important contributions is the concept of utility, which makes the old solutions and strategies plausible once again. The utility function was fashioned precisely for this purpose; and in this chapter we will discuss utility functions and their role in game theory.

The heart of the difficulty is the ambiguity of zero-sum.

Taking the simplest view, a game is zero-sum if it satisfies
a certain conservation law: a game is zero-sum if, during
the course of the game, wealth is neither created nor destroyed. In this sense, ordinary parlor games are zero-sum.

But this definition won't do, for it is not the payoffs in
money that are important. Throughout the discussion of
the two-person, zero-sum game, it was assumed that each
player was doing his best to hurt his opponent; if this assumption fails, the rest of the theory fails with it. What is
needed to make our earlier argument plausible is the assurance that the players will indeed compete. A parent playing a card game with a child for pennies is playing a zero-
sum game in our original sense, yet the arguments of the
last chapter do not apply to it. Of course, if we *assume*, as
we did earlier, that a player's goal is to maximize his expected payoff in money, there is no problem, but this begs
the question; it is the validity of this assumption that is
in doubt.

As a matter of fact, there are many situations in which
people *don't* act so as to maximize their expected winnings.
This has nothing to do with the formal theory; it is an observation about life. The game we used as an illustration at
the end of the last chapter is only one example in which
the assumption is incorrect; there are many others. Let us
consider just a few from everyday life.

There are many games that have a negative average payoff which nevertheless attract a large number of players.
For instance, people buy lottery tickets, play the "numbers
game," bet at the race track, and gamble in Las Vegas and
Reno. This isn't to say that people bet aimlessly without
pattern or purpose; certain bets are generally preferred to
others. In studies of parimutuel betting at the race track,
for instance, it was observed that bettors consistently gave
more favorable odds on long-shots than experience war-
ranted, while the favorites, at more attractive odds, were
neglected. Just what it is that makes one bet more attractive
than another is not always clear; what is clear is that the

players are trying to do something other than maximize their average winnings.

It isn't just the gamblers who are willing to play games with unfavorable odds; such games are also played by conservatives who wish to avoid large swings. Commodity-futures markets are created because farmers at planting want to be sure prices don't drop too low by harvest. Also, almost every adult carries personal insurance of one sort or another, yet the "game" of insurance has a negative value, for the premiums cover not only the benefits of all policyholders but insurance company's overhead and commissions as well. Incidentally, insurance is just a lottery turned around. In both cases, the player puts in a small amount; the lottery player has a small chance to win a fortune, and the insurance policyholder avoids the small chance of having a catastrophe befall him.

The prevalence of insurance policies reflects a willingness to pay a price for security; this aversion to risk has also been observed in the laboratory.

H. Markowitz asked a group of middle-class people whether they would prefer to have a smaller amount of money with certainty or an even chance of getting ten times that much. The answers he received depended on the amount of money involved. When only a dollar was offered, all of them gambled for ten, but most settled for a thousand dollars rather than try for ten thousand, and all opted for a sure million dollars.

It should be emphasized that the failure to maximize average winnings is not simply an aberration caused by a lack of insight. In 1959, Alvin Scodel, Philburn Ratoosh, and J. Sayer Minas asked the subjects of an experiment to select one of several bets. There was no correlation whatever between the bets selected and the intelligence of the subject. Graduate students in mathematics were included among the subjects, and they certainly could do the necessary calculations if they wished.

So if the theory is to be realistic—and if it is not realistic, it is nothing—then we cannot assume that people are concerned only with their average winnings. In fact, we cannot make any general assumption about people's wants, because different people want different things. What is needed is a mechanism that relates the goals of a player, whatever they are, to the behavior that will enable him to reach these goals. In short, a theory of utility.

Before you can make sensible decisions in a game, both the goals of the players and the formal structure of the game must be taken into account. In this process of decision-making, there is a natural division of labor. The game-theorist, to paraphrase Lewis Carroll, must pick the proper road after he is told the player's destination; and the player doesn't have to know anything about game theory, but he has to know what he likes—a slightly altered version of the popular bromide.

The problem is to find a way for the player to convey his attitudes in a form that is useful to the decision-maker. Statements such as "I detest getting caught in the rain" or "I love picnics" do not help someone decide whether to call off a picnic when the weather prediction is an even chance of rain. There is no hope of completely describing subjective feelings quantitatively, of course, but, using utility theory, it is possible to convey enough of these feelings (under certain conditions) to satisfy our present purpose.

Utility Functions: What They Are, How They Work

A utility function is simply a "quantification" of a person's preferences with respect to certain objects. Suppose we are concerned with three pieces of fruit: an orange, an apple, and a pear. The utility function first associates with each piece of fruit a number that reflects its attractiveness. If the pear was desired most and the apple least, the utility of the pear would be greatest and the apple's utility would be least.

The utility function not only assigns numbers to fruit; it assigns numbers to lotteries that have fruit as their prizes. A lottery in which there is a 50 percent chance of winning an apple and a 50 percent chance of winning a pear might be assigned a utility of 6. If the utilities of an apple, an orange, and a pear were 4, 6, and 8, respectively, the utilities would reflect the fact that the person was indifferent (had no preference) between a lottery ticket and an orange, that he preferred a pear to any other piece of fruit or to a ticket to the lottery, and that he preferred a ticket to the lottery or any other piece of fruit to an apple.

Also, utility functions assign numbers to all lotteries that have as prizes tickets to other lotteries; and each of the new lotteries may have as its prizes tickets to still other lotteries, so long as the ultimate prizes are pieces of fruit.

This is still not enough, however. There is one more thing that von Neumann and Morgenstern demand of *their* utility functions which make them ideally suited for their theory. The utility functions must be so arranged that the utility of any lottery is always equal to the weighted average utility of its prizes. If in a lottery there is a 50 percent chance of winning an apple (which has a utility of 4), and a 25 percent chance of winning either an orange or a pear (with utilities 6 and 8, respectively), the utility of the lottery would necessarily be 5 1/2.

The Existence and Uniqueness of Utility Functions

It is easy enough to list conditions that one would like the utility functions to satisfy; it is quite another thing to find a utility function that does. Given a person of arbitrary tastes, is it always possible to find a utility function that reflects them? Can these tastes be reflected by two different utility functions?

Taking the second question first, the answer is yes. Once a utility function is established, another can be obtained from it by simply doubling the utility of everything; and still another, by adding one to the utility of everything. In fact, if we take any utility function and multiply the utility of everything by any positive number (or add the same number to every utility), a new utility function is obtained which works just as well.

It is also possible that there is no utility function that will do; this would be the case if the person's tastes lacked "internal consistency." Suppose a person, when given a choice between an apple and an orange, prefers the apple. That would mean that the utility of an apple would have to be greater than the utility of an orange. If, between an orange and a pear, he prefers the orange, the utility of an orange would have to be greater than that of a pear. Since the utilities are ordinary numbers, it follows that the utility of an apple is necessarily greater than the utility of a pear. If, when given the choice between an apple and a pear, the person chooses the pear, establishing a utility function would be hopeless. *There is no way one can assign numbers to the three pieces of fruit that will simultaneously reflect all three of these preferences.* There is a technical name for what we have just described: the player's preferences are said to be *intransitive*.

If the preferences of a player are sufficiently consistent— that is, if they satisfy certain requirements—the preferences can be expressed concisely in the form of a utility function. The precise requirements can be stated in many ways; we have included in Appendix A to this chapter a version that was originally suggested by Duncan R. Luce and Howard Raiffa.

Henceforth, we will suppose that the preferences of each player are expressed by a utility function and that the payoffs are given in "utiles"—the units in which utility functions are expressed. When we speak of a zero-sum game now, we mean a game in which the sum of all the payoffs (in utiles) is always zero. When two players are involved in a zero-sum game in this new sense, their interests must necessarily be opposed.

The advantage of this new definition of zero-sum is that it justifies the work we did earlier on two-person, zero-sum games, if applied to the new zero-sum games. But, admittedly, there is a corresponding disadvantage in the new approach. This might have been anticipated as a consequence of our quasi-conservation law: games that are easy to analyze don't come up very often. The difficulty is that this new type of zero-sum game is rare, and hard to recognize. Before we changed the definition, you only had to know the rules of a game such as poker to see immediately that it was zero-sum. Now the problem of recognition is much more subtle and involves subjective factors such as the players' attitude toward risk. This makes the theory much more troublesome to apply.

Some Potential Pitfalls

Utility theory is easily misunderstood. The reason for this is partially historical: "utilities" have been in existence for some time, but the term has not always been used in a consistent or clear way. In an article called "Deterrence and Power," Glenn Snyder, in the *Journal of Conflict Resolution* in 1960, gave a number of good examples of how one can go wrong, as follows:

"The figures which follow [i.e., the payoff matrix] are based on the assumptions that both sides are able to translate all relevant values into a 'common denominator' 'utility,' that they can and do estimate probabilities for each other's moves, and that they act according to the principle of 'mathematical expectations.' The latter states that the 'expected value' of any decision or act is the sum of the expected values for all possible outcomes, the expected value for each outcome being determined by multiplying its value to the decision-making unit times the probability that it will occur. To act 'rationally' according to this criterion means simply to choose from among the available courses of action the one which promises to maximize expected value (or minimize expected cost) over the long run. There are reasons—apart from the practical difficulty of assigning numerical values to the elements involved—why the mathematical expectation criterion is not entirely suitable as a guide to rationality in deterrence and national security policy. However, it is useful as a first approximation, the necessary qualifications—having to do with the problem of uncertainty and the disutility of large losses—will be made presently."

Snyder starts out all right; you do have to "translate all relevant values into a 'common denominator' 'utility.'" That this is possible must be assumed, but that is *all* you have to assume. Snyder is also willing to assume that "they act according to the principle of 'mathematical expectations.'" But why should a person, or a country, want to maximize its average utility, any more than it would want to maximize its average winnings in money?

Both the last question and Snyder's assumption put the cart before the horse. The person's wants come first; the utility function, if it exists, comes second. The person is *not* trying to maximize his utility—average or otherwise; very likely, he doesn't even know such a thing exists. A player may act *as though* he were maximizing his utility function, but this is not because he means to but because of the judicious way the utility function was established. What happens, at least in theory, is that the preferences of the player are observed and then a utility function is established which the player *seems* to be maximizing.

Snyder's statement: "To act 'rationally' according to this criterion means simply to choose from among the available courses of action the one which promises to maximize expected value (or minimize expected cost) over the long run" can be criticized on two counts. For one thing, the problem is put the wrong way around, as it was earlier. You don't maximize your utility because you're rational; your preferences are observed *first* and *then* the utility function is established. Moreover, "over the long run" has nothing to do with it. Utilities are assigned to one-shot gambles. If a person takes the view that life is a series of gambles in which good and bad luck cancel each other, he may decide to maximize his average return in dollars; and this is all very well. But if he prefers less risky alternatives, that's all right too, even if it lowers his expected return. Utility theory will accommodate either attitude. In the deterrence "game," gambles are taken without any assurance that they will recur; in fact, there is good reason to believe they will not.

And, finally, the author's reservations about the "disutility of large losses" entirely misses the point. Either a utility function exists or it doesn't. If the utility function associates with large losses numbers that are inaccurate, then it isn't a utility function at all. A utility function, if it does anything, must reflect a person's preferences accurately; it serves no other purpose.

Constructing a Utility Function

Granted that there is a need for a utility function, there still remains the problem of actually determining it. Roughly speaking, this is how it can be done. The person is asked to make many simple choices between two things; it may be pieces of fruit, or lottery tickets. He might be asked, for example, whether he would prefer a ticket to a lottery in which there is a 3/4 chance of winning a pear and a 1/4 chance of winning an apple, or the certainty of getting an orange. On each question he must indicate a preference for one of the alternatives or state that he is indifferent to both. It is possible, on the basis of these simple choices, to establish a single utility function that assigns a number to every piece of fruit and every possible lottery—reflecting all the preferences of an individual simultaneously, providing his preferences are consistent. In effect, the fruit and lotteries are put on a single dimension in which every piece of fruit and every lottery ticket can be simultaneously compared.

It should be understood that in the process of establishing utility function, nothing new was added; the grand final ordering is implicit in the simple choices made earlier. But the practical advantage of having a concise utility function rather than a great many individual preferences is enormous.

Are People's Preference Patterns Really Consistent?

So far we have assumed that people for whom utility functions are to be established have consistent preferences; that is, preferences that satisfy the six consistency conditions listed in Appendix A to this chapter. On the face of it, these conditions seem intuitively appealing, and one would think that most people would accept them with few if any objections. It has even been suggested that these conditions —or, at any rate, some of them—should be used as a *definition* of rationality in decision-making. But it turns out that people often have, or seem to have, inconsistent attitudes that preclude the construction of a utility function. Let's look at some of the things that can go wrong.

A fair amount of effort has been expended to study betting behavior experimentally and, in particular, to see whether betting behavior is consistent in the sense mentioned earlier. In a number of experiments, Ward Edwards compared the bets that subjects chose with certain other variables, such as the average amount the subject could win, the state of the subject's fortunes when he made the choice, the actual numerical probabilities involved, etc. He found, among other things, that certain subjects tended to make bets in which the probabilities involved were (1/2, 1/2) rather than (1/4, 3/4), even though the average return was the same in both cases. It is easy to show, and a proof is included in Appendix B to this chapter, that a person cannot consistently prefer the one type of bet to the other and still satisfy the six conditions necessary for consistency.

Another source of difficulty is the variation in people's preferences over a period of time. This might not appear to present too serious a problem, since variations would occur gradually; but that is not the case. The interaction between what happens in the game and the attitudes of the players cannot be neglected at any time. A worker who has been out on strike for a long time looks at an offer with different eyes from a worker just starting to negotiate; and, during the process of negotiation, flexible attitudes may harden and what were once acceptable possibilities may become unacceptable.

Anyone who has watched people gambling will no doubt have observed the same thing. In a poker game, for example, as time passes, the bets grow larger: apparently the amount that one considers an acceptable risk increases as the game proceeds. The same phenomenon has been observed experimentally by comparing the size of bets made in the early and the late races at the track. A subjective but graphic picture of what happens is given by Dostoevsky in his novel *The Gambler*:

"As for me, I lost every farthing quickly. I staked straight off twenty friedrichs d'or on even and won, staked again and again won, and went on like that two or three times. I imagine I must have had about four hundred friedrichs d'or in my hands in about five minutes. At that point I ought to have gone away, but a strange sensation rose up in me, a sort of defiance of faith, a desire to challenge, to put out my tongue at it. I laid down the largest stake allowed—four thousand gulden—and lost it. Then getting hot, I pulled out all I had left, staked it on the same number, and lost again, after which I walked away from the table as though

I were stunned, I could not even tell what happened to me. . . ."

There are many other problems as well. Experiments have shown that decisions often depend on seemingly irrelevant variables. People make one type of bet when playing for money and another when playing for wooden chips that are worth money. They bet one way when others are present and another way when they're alone. Their history —the success they've had in the game so far—influences their attitude toward risk. People select inconsistently; they pick one of two bets one time, and the other the next. And their preferences are intransitive: they prefer A when offered a choice between A and B, and B when offered a choice between B and C. And when offered a choice between A and C, they prefer C.

Note how the context of a problem can influence decision-making: "A man buying a car for $2,134.56 is tempted to order it with a radio installed, which will bring the total price to $2,228.41, feeling that the difference is trifling. But when he reflects that, if he already had the car, he would certainly not spend $93.85 for a radio for it, he realizes that he has made an error" [from *The Foundations of Statistics*, by Leonard J. Savage].

Despite all the difficulties—despite the seeming irrationality of people's behavior, despite the inconsistencies—utility functions have been established successfully. In one case, experimenters observed the bets that subjects chose when they were offered simple alternatives. On the basis of these observations, they were able to predict what the subjects would do when presented with much more complicated decisions. How was this possible, in view of all the objections mentioned?

Not all the objections can be explained away, of course, but the problem is not as serious as it may seem. Take the case of intransitive preferences, for instance. Some experimenters feel that true intransitivities rarely arise. What happens, they say, is that people are forced to choose between alternatives to which they are indifferent. The answer, then, is dependent on momentary whim. If a person has difficulty deciding whether to buy a car for $2,000, it should not be surprising if one moment he refuses to buy at a dollar less and the next moment agrees to buy at a dollar more. In several experiments, it was observed that subjects who make intransitive choices don't make them consistently. In one experiment, for instance, subjects were asked to list certain crimes in the order of their seriousness and to determine the severity with which they should be punished. While occasional intransitivities turned up, when the task was repeated they almost invariably disappeared.

It is likely that much "irrational behavior" can be avoided by identifying and controlling the significant variables. If, for example, people bet more aggressively when they are with others than when they are alone (and they do), this must be taken into account. Also, the subjects should have some experience in decision-making, so they can take into consideration the consequences of their actions. In Savage's example (buying a car with a radio), this is precisely what was needed.

Finally, people must be properly motivated. In some experiments, subjects stated that they had varied from what they believed to be the most prudent course, just to keep things exciting. This is a serious problem, since as a rule the experimenter has limited resources and cannot pay enough to motivate the players to the degree he would like. People betting in a "hot, humid, exciting, and highly disordered carnival setting," as one experimenter put it, behaved quite differently from bettors in an artificial, insulated, experimental environment.

Appendix A

Six Conditions That Guarantee the Existence of a Utility Function

If a player's preferences are to be expressed by a utility function, these preferences must be consistent; that is, they must satisfy certain conditions. These conditions may be expressed in several more or less equivalent ways; we have used the formulation suggested by Luce and Raiffa. For convenience, the word "object" will represent either a piece of fruit or a lottery.

1. *Everything is comparable.* Given any two objects, the player must prefer one to the other or be indifferent to both; no two objects are incomparable.

2. *Preference and indifference are transitive.* Suppose A, B, and C are three different objects. If the player prefers A to B and B to C, he will prefer A to C. If the player is indifferent between A and B, and B and C, he will be indifferent between A and C.

3. *A player is indifferent when equivalent prizes are substituted in a lottery.* Suppose in a lottery one prize is substituted for another but the lottery is left otherwise unchanged. If the player is indifferent between the old and new prizes, he will be indifferent between the lotteries. If the player prefers one prize to the other, he will prefer the lottery that offers the prize he prefers.

4. *A player will always gamble if the odds are good enough.* Suppose that, of three objects, A is preferred to B and B is preferred to C. Consider the lottery in which there is a probability p of getting A and a probability $(1-p)$ of getting C. Notice that if p is zero the lottery is equivalent to C, and if p is 1, the lottery is equivalent to A. In the first case the lottery is preferable to B, while in the second case

B is preferable to the lottery. Condition 4 states that there is a value for *p* between zero and 1 which will make the player indifferent between *B* and the lottery.

5. *The more likely the preferred prize, the better the lottery.* In lotteries *I* and *II* there are two possible prizes: objects *A* and *B*. In lottery *I*, the chance of getting *A* is *p*; in lottery *II*, the chance of getting *A* is *q*. *A* is preferred to *B*. Condition 5 requires that if *p* is bigger than *q*, lottery *I* be preferred to lottery *II*; and, conversely, if lottery *I* is preferred to lottery *II*, *p* is greater than *q*.

6. *Players are indifferent to gambling.* A player's attitude toward a compound lottery—a lottery in which the prizes may be tickets to other lotteries—is dependent only on the ultimate prizes and the chance of getting them as determined by the laws of probability; the actual gambling mechanism is irrelevant.

Appendix B

Bets of Equal Average Value Must Be Equally Preferred

It has been observed during experiments that subjects are partial to bets involving certain probabilities. Specifically, it was found that certain people preferred *any* bet in which they obtained one of two amounts of money with probability 1/2, to a bet in which the probabilities are 1/4 and 3/4, providing the average value obtained was the same. No utility function of the type we have been considering can possibly describe such preferences. To see this, let us suppose the contrary—that each amount of money has associated with it a certain utility. Let us denote the utility of $X by $U(X)$.

Consider two bets. In bet 1, there is an even chance of getting $150 and an even chance of getting $100; in bet 2, there is 1/4 of a chance of getting $200 and a 3/4 chance of getting $100. In both bets, the average return is $125, so, by our earlier assumption, the first bet is preferred. In terms of utilities:

1/2 $U(100)$ + 1/2 $U(150)$ > 1/4 $U(200)$ + 3/4 $U(100)$; or alternately:

(i) 2U(150) > U(100) + U(200)

If the amounts in bet 1 are changed to $50 and $200, and the amounts in bet 2 are changed to $50 and $150, we have:

(ii) 1/2 U(50) + 1/2 U(200) > 1/4 U(50) + 3/4 U(150)

Finally, changing the amounts in bet 1 to $50 and $100, and the amounts in bet 2 to $150 and $50, we get:

(iii) 1/2 U(50) + 1/2 U(100) > 3/4 U(50) + 1/4 U(150)

Adding *(ii)* and *(iii)* and multiplying by 2, we get:

$U(200)$ + $U(100)$ > 2 $U(150)$, which contradicts *(i)*

5

The Two-Person, Non-Zero-Sum Game

The theory of two-person, zero-sum games is unusual in that it enables one to find solutions—solutions, moreover, that are universally accepted. In this respect, zero-sum games are unlike the actual problems that arise in everyday life, which generally do not lend themselves to straightforward answers. For non-zero-sum games we cannot do nearly as well. In most games of any complexity, there is no universally accepted solution; that is, there is no single strategy that is clearly preferable to the others, nor is there a single, clear-cut, predictable outcome. As a rule, we will have to be content with something less than the unequivocal solutions we obtained for zero-sum games.

For convenience, let us regard all two-person games as lying in a continuum, with the zero-sum games at one ex-

treme. In a two-person game, there are generally both competitive and cooperative elements: the interests of the players are opposed in some respects and complementary in
others. In the zero-sum game, the players have no common
interests. In the completely cooperative game at the other
extreme, the players have nothing but common interests.
A pilot of an airplane and the operator in the control tower
are engaged in a cooperative game in which they share a
single common goal, a safe landing. Two sailboats maneu-
vering to avoid a collision, and two partners on a dance
floor, are also playing a cooperative game. The problem in
such a game is easy to solve, at least conceptually: it consists of coordinating the efforts of the two players efficiently
(by means of dancing lessons, for instance).

   The rest of the two-person games, and the ones with which
we will be primarily concerned, fall between these two
extremes. Games with both cooperative and competitive elements are generally more complex, more interesting, and
encountered more frequently in everyday life than pure
competitive or cooperative games. Some examples of these
games are: an automobile salesman negotiating with a customer (both want to consummate the sale, but they differ
on the price), two banks discussing a merger, two competing stores, etc. In each of these games the players have
mixed motives. Also, there are many situations where the
parties *seem* to have no common interests but really do.
Two nations at war may still honor a cease-fire, not use
poison gas, and refrain from using nuclear weapons. In fact,
zero-sum games are almost always an approximation to
reality—an ideal that is never quite realized in practice. In
the previous chapter, for example, if the examples are varied
even slightly, they cease to be zero-sum. In the marketing
problem, we assumed that total consumption was constant.
But if, when the advertising schedule was changed, total
consumption also changed (as in practice it would), the
game would be non-zero-sum. And even if the assumption
were valid but the firms conspired together to reduce the

amount of advertising (and thus lower costs), the game would also be non-zero-sum.

Analyzing a Two-Person, Non-Zero-Sum Game

The easiest way to get some insight into the non-zero-sum game is to try to analyze one in the same way that we analyzed the zero-sum game. Suppose we start with the matrix shown in Figure 15. Notice that in non-zero-sum games it is necessary to give the payoffs for *both* players, since the payoff of one player can no longer be deduced from the payoff of the other, as it can in zero-sum games. The first number in the parentheses is the payoff to player *I*, and the second number is the payoff to player *II*.

Player *II*

		A	B
	a	(0,0)	(10,5)
Player *I*	b	(5,10)	(0,0)

Figure 15

It is obvious what the players' common interest is: both must avoid the zero payoffs. But there still remains the problem of determining who will get 5 and who will get 10. One way to attack the problem, a way that was successful before, is to look at the game from the point of view of one of the players—say, player *I*. As player *I* sees it, the game has the matrix shown in Figure 16. Since the payoffs to player *II* don't concern him directly, they are omitted.

Ultimately, player *I* must decide what he wants from this game and how he should go about getting it. A good way to start is by determining what he can get without any help from player *II*; this, at any rate, is the *least* he should be willing to settle for.

Using the techniques of the game, player *I* will find that, by *playing strategy* a 1/3 *of the time, he can get an average return of* 10/3. *If player* II *plays strategy* A 2/3 *of the time,* that's all he can get.

Player *II*

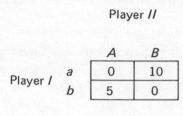

Figure 16

If player *II* makes the same sort of calculation, using his own payoff matrix and ignoring the payoff to player *I, he too can get* 10/3 *by playing strategy* A 1/3 *of the time.* If *player I plays strategy* a 2/3 *of the time, player* II will not get any more.

So far, so good. It looks as if we are in much the same position as we were in the zero-sum game: each player can get 10/3; each player can be stopped from getting any more. Why not, then, call this the solution?

The trouble with this "solution" is that the payoffs are too low. If the players can manage to get together on either one of the non-zero payoffs, they will *both* do better. The argument we used to support the (10/3, 10/3) payoff, which was sound enough in the zero-sum game, fails here. *Although each player can prevent the other from getting more than* 10/3, *there is no reason why he should.* It is no longer true that a player can get rich only by keeping his opponent poor. Although it is true that the players cannot be prevented from getting 10/3, they are foolish if they don't aspire to something better.

But how to go about doing better? Suppose player *I* anticipates everything that we have said so far and guesses that player *II* will play strategy A 1/3 of the time, to ensure a payoff of 10/3. If player *I* switches to the pure strategy a, player *II* will still get his 10/3, but player *I* will now get 20/3: twice as much as before. The trouble with this argument is that player *II* may get ambitious also. He, too, may decide to double his return and, anticipating conservative play by player *I*, play the pure strategy A. If they both do this simultaneously, they will both get nothing. As in the zero-sum game, this sort of circular reasoning does not take you very far.

Up to now, our main purpose has been to give some idea of the kind of problems that may arise. To this end, and also for future reference, let us look at some examples.

The Gasoline Price War

Two competing gas stations buy gas at 20 cents a gallon and, together, sell a thousand gallons a day. They have each been selling gas at 25 cents a gallon and dividing the market evenly. One of the owners is thinking of undercutting the other. The prices, which are quoted in whole numbers, are set independently by each station in the morning, and remain fixed for the day. There is no question of customer loyalty, so a lower price at either station attracts virtually all the sales. What price should each station owner set? If one owner knows the other is going out of business the next day, should that make a difference? What if he is going out of business ten days later? What if he is going out of business at some indefinite time in the future?

For each set of prices set by the gas stations, corresponding profits are entered in the matrix in Figure 17. If gas station *I* charges 24 cents and gas station *II* charges 23 cents, gas station *II* captures the entire market; he makes 3 cents profit per gallon on a thousand gallons, or $30 profit. Gas station *I* makes no profit at all.

Gas Station *II*'s
Price per Gallon

			25¢	24¢	23¢	22¢	21¢
		25¢	(25,25)	(0,40)	(0,30)	(0,20)	(0,10)
Gas	Price	24¢	(40,0)	(20,20)	(0,30)	(0,20)	(0,10)
Station *I*'s	per	23¢	(30,0)	(30,0)	(15,15)	(0,20)	(0,10)
	Gallon	22¢	(20,0)	(20,0)	(20,0)	(10,10)	(0,10)
		21¢	(10,0)	(10,0)	(10,0)	(10,0)	(5,5)

Figure 17

Let us start by trying to answer the easiest question: what price should a gas station set if its competitor is going out of business the next day? The first thing to note is that one should *never* charge less than 21 cents or more than 25 cents. When the price drops below 21 cents, there is no profit, and when it rises above 26 cents, you lose your market. Note also that you should never charge 25 cents, since this price is dominated by 24 cents. If your competitor charges either 24 or 25 cents, you do better by charging 24 cents, and if he charges something else, you don't do any worse. A gas station should not charge 25 cents, then, and he may reasonably assume that his competitor won't either.

Once 25 cents is ruled out as a selling price for *either* gas station, one can, by the same reasoning, rule out 24 cents as a selling price, because it is dominated by a price of 23 cents. As a matter of fact, one can successively rule out every price but 21 cents.

There is a paradox here. Starting from the status quo, where the stations charged 25 cents a gallon and received $25 profit apiece, they managed, by reasoning logically, to get to a position where they charge 21 cents and receive only $5 profit apiece.

If one of the gas stations is known to be going out of business in ten days rather than in one day, it's a bit more complicated. The old argument is all right as far as it goes, but now we are concerned with the profits not only of the next day but of the following nine as well. And if a player drops his price one day, he can be sure that his competitor will do the same on the next. It turns out, however, that even here an apparently airtight argument can be made for charging 21 cents. It is only when the competition continues indefinitely that the argument finally loses force.

This game is a variation of the "prisoner's dilemma"; we will say more about it later.

A Political Example

The state legislature is about to vote on two bills that authorize the construction of new roads in cities A and B. If the two cities join forces, they can muster enough political power to pass the bills, but neither city can do it alone. If a bill is passed, it will cost the taxpayers of both cities a million dollars and the city in which the roads are constructed will gain $10 million. The legislators vote on both bills simultaneously and secretly; each legislator must act on each bill without knowing what anyone else has done. How should the legislators from cities A and B vote?

The payoff matrix for this game is shown below. Since a city always supports its own bond issue, the legislators have only two strategies: supporting, or not supporting, the sister city.

The entries in the matrix in Figure 18 are in millions of dollars. As an example of how they were computed, suppose city A supports city B's bond issue, but B doesn't support A's. Then one bond issue passes, each city pays a million dollars, B gets $10 million, and A gets nothing. The net effect is that A loses a million dollars and B gets $9 million.

The political example is really a "prisoner's dilemma" in a different form. As before, the smartest thing seems to vote against the other city's bond issue. And, just as before, if both cities follow this strategy, they will both get nothing, rather than getting the $8 million apiece they could have gotten otherwise.

City *B*

		Support *A*'s Bond Issue	Withhold Support from *A*'s Bond Issue
City *A*	Support *B*'s Bond Issue	(8,8)	(−1,9)
	Withhold Support from *B*'s Bond Issue	(9, −1)	(0,0)

Figure 18

The Battle of the Sexes*

A man and his wife have decided that they will go either to the ballet or to the prize fights that evening. Both prefer to go together rather than going alone. While the man would most prefer to go with his wife to the fights, he would

Man

		Fights	Ballet
Wife	Fights	(2,3)	(1,1)
	Ballet	(1,1)	(3,2)

Figure 19

prefer to go with her to the ballet rather than go to the fights alone. Similarly, his wife's first choice is that they go together to the ballet, but she too would prefer that they go to the fights together rather than go her own way alone. The matrix representing this game is shown in Figure 19.

* From *Games and Decisions,* by R. Duncan Luce and Howard Raiffa.

The payoffs reflect the order of the players' preferences. This is essentially the same game we talked about at the opening of this chapter.

A Business Partnership

A builder and an architect are jointly offered a contract to design and construct a building. They are offered a single lump sum as payment and given a week to decide whether they will accept. The approval of both is required, so they must decide how the lump sum will be divided. The architect writes to the builder, suggesting that the profits be divided equally. The builder replies, committing himself to the contract in writing, provided he receives 60 percent of the profits. He informs the architect, moreover, that he is going on a two-week vacation, that he cannot be reached, and it is up to the architect to accept the contract on these terms or reject it entirely.

The architect feels he is being exploited. He feels that his services are as valuable as the builder's and that each should receive half the profits. On the other hand, the probable profit is large, and under different circumstances, he would consider 40 percent of such a total an acceptable fee. Should he accept the builder's offer? Note that the builder has no more options; only the architect has a choice —to accept or reject the contract. This is reflected in the payoff matrix shown in Figure 20.

Builder

Architect	Accept Offer	(40%, 60%)
	Reject Offer	(0,0)

Figure 20

A Marketing Example

Each of two firms is about to buy an hour of television time to advertise their competitive products. They can advertise either in the morning or in the evening. The television audience is divided into two groups. Forty percent of the people watch during the morning hours, and the remaining 60 percent watch in the evening; there is no overlap between the groups. If both companies advertise during the same period, each will sell to 30 percent of the audience watching and make no sales to the audience not watching. If they advertise during different periods, each will sell to 50 percent of the audience watching. When should they advertise? Should they consult before making a decision? The payoff matrix for the game is shown in Figure 21.

Firm *II*

		Advertise in Morning	Advertise in Evening
Firm *I*	Advertise in Morning	(12,12)	(20,30)
	Advertise in Evening	(30,20)	(18,18)

Figure 21

The matrix entries represent the percentage of the total audience that each firm captures. If firm *I* picks an evening hour and firm *II* picks a morning hour, firm *I* sells to half of 60 percent, or 30 percent, and firm *II* sells to half of 40 percent, or 20 percent.

Some Complications

Two-person, zero-sum games come up in many different contexts, but they always have the same basic structure. By looking at the payoff matrix, one can pretty well tell "the whole story." This is not the case in non-zero-sum games. Besides the payoff matrix, there are many "rules of the game" that markedly affect the character of the game, and these rules must be spelled out before one can talk about the game intelligently. It is impossible to say very much on the basis of the payoff matrix alone.

To see this more clearly, go back to the first example discussed in this chapter—in which the payoffs were (0, 0), (10, 5), and (5, 10). There were several details that we deliberately omitted but which are obviously important. Can the players consult beforehand and agree to their respective strategies in advance? Are these agreements binding? That is, will the referee or whoever enforces the rules insist that the agreement be carried out, or does the agreement have moral force only? Is it possible for the players, after one has received 5 and the other 10, to make payments to each other so they both receive 7 1/2? (In some games, they can; in others, they can't.) In the gasoline price war, is it feasible (legal) for the two businesses to conspire to fix prices?

It is to be expected that these (and other) factors will have a strong influence on the outcome of the game; but their actual effect is often different from what one would expect.

After one studies the two-person, zero-sum game, certain aspects of the non-zero-sum game seem like something out of *Alice in Wonderland*. Many "obvious" truths—fixed stars in the firmament of the zero-sum game—are no longer valid. Look at a few:

1. One would think that the ability to communicate could never work to a player's disadvantage. After all, even if

a player has the option to communicate, he can refuse to exercise it and so place himself in the same situation as if facilities for communication were nonexistent; or so it would seem. The facts are otherwise, however. The *inability* to communicate may well work to one player's advantage, and this advantage is lost if there is a way to communicate, even though no actual communication occurs. (In the zero-sum game, this doesn't come up. There, the ability to communicate is neither an advantage nor a disadvantage, since the players have nothing to say to each other.)

2. Suppose that in a symmetric game—a game in which the payoff matrix looks exactly the same from both players' point of view—player *I* selects a strategy first. Player *II* picks his strategy after he sees what player *I* has done. One would think that in a game where the players have identical roles except that player *II* has the benefit of some additional information, player *II*'s position would be at least as good as player *I*'s. In the zero-sum game, this situation could not possibly be to player *I*'s advantage. In the non-zero-sum game, however, it may. Let us carry the point a little further. It is often to a player's advantage to play before his opponent, even if the rules do not require that he do so, or, alternately, to announce his strategy so that his decision becomes irrevocable.

3. Suppose that the rules of a game are modified so that one of the players can no longer pick some of the strategies that were originally available to him. In the zero-sum game, the player might not lose anything, but he certainly could not gain. In the non-zero-sum game, he might very well gain.

4. It often happens in a non-zero-sum game that a player gains if his opponent does not know his utility function. This is not surprising. What is surprising is that at times it is an advantage to have your opponent know your utility function, and he may be worse off when he learns it. This doesn't come up in zero-sum games: there it is assumed that each player knows the other's utility function.

Communication

The extent to which players can communicate has a profound effect on the outcome of a game. There is a wide spectrum of possibilities here. At one extreme we have games in which there is no communication whatever between the players and the game is played only once (later we will discuss why this matters). At the other extreme are the games in which the players can communicate freely. Generally, the more cooperative the game—the more the players' interests coincide—the more significant is the ability to communicate. In the zero-sum game which is completely competitive, communication plays no role at all. In the completely cooperative game, the problem is solely one of communication, and so the ability to communicate is crucial.

Cooperative games in which the players can communicate freely present no conceptual difficulty. There may be technical difficulties, of course, such as exist when a control tower directs a pilot in heavy traffic. But two players who can't communicate directly, such as two sailboat captains trying to avoid a collision in choppy waters or two guerrillas behind enemy lines, have a problem.

Professor Thomas C. Schelling feels that certain tactics are useful in this kind of game. He suggests that players observe their partners' behavior for clues to what they will do later; this is more or less what happens on the dance floor. Generally accepted cultural norms may also serve the purpose. In England, for instance, you drive on the left if you want to avoid collisions, and you assume others will also, whereas in the United States you drive on the right. Also, it may be possible to exploit such prominent features as may exist in the game. If two people who are to meet in New York City at a certain time have no way of communicating, or any established pattern on which to rely, they may try to meet at some prominent landmark such as

Grand Central Station. This last principle can be applied
to the game shown in Figure 22.

Player *II*

	(1,1)	(0,0)	(0,0)
Player *I*	(0,0)	(1,1)	(0,0)
	(0,0)	(0,0)	(1,1)

Figure 22

The prominent feature of the game is, obviously, its
symmetry, and it seems perfectly clear that both players
should play a middle strategy if they want to get together.
But, in fact, playing the middle strategy is justified only if
the other player plays it too. And that is questionable, as
we shall see when we discuss experimental games.

In completely cooperative games, communication is an
unalloyed blessing. In games in which the players have some
conflicting interests, communication has a more complex
role. To see this, look at the game shown in Figure 23. This
is one form of the "prisoner's dilemma." There are two basic
properties of this game that are worth noting. No matter
which strategy his partner plays, *a player always does best
if he plays strategy* B. Also, no matter what strategy a player
finally selects, he will *always do better* if his *partner plays
strategy* A. Putting these two together, we have a conflict:
on the one hand, each player should play strategy *B*; on
the other hand, each is hoping the other will play strategy
A. It is to the advantage of both, moreover, that the conflict
be resolved by both choosing A rather than B. How do the
players go about getting this result?

If the game is played only once, the prospects for the
(5, 5) outcome—sometimes called the cooperative outcome—
are pretty bleak. Neither player can influence his partner's
play, so the best each can do is play B. When the game is

played repeatedly, it is a little different; it may be possible then to play in such a way that one's partner is induced to play *A*. And eventually it may be possible to get (5, 5) payoff.

Player *II*

		A	B
Player *I*	A	(5,5)	(0,6)
	B	(6,0)	(1,1)

Figure 23

Granted there is an opportunity to arrive at this payoff when the game is played repeatedly, how do you invite your partner to cooperate in a specific situation? One way to do it is unilaterally to adopt an apparently inferior strategy (such as strategy *A* in our illustration) and hope the other player catches on. If your partner turns out to be either stupid or stubborn, there is nothing to do but revert to the cutthroat strategy, *B*, and settle for a payoff of 1. Thus, in fact, it may be possible to communicate in a game even if there is no direct contact between the players.

This kind of tacit communication is generally ineffective, however, most likely because it is generally misunderstood. Merrill M. Flood described an experiment in which a game, similar to the one we discussed earlier, was played a hundred times by two sophisticated players. The players made written notes during the game, and these left little doubt that the players were not communicating. This same conclusion was also reached in other experiments, as we will see later.

Even if the message requesting cooperation does get through, it may not be in the best interests of the receiver to accept it immediately. It has been suggested that the appropriate response is to "misunderstand" at first, and let the other player "teach" you; then, before he loses hope, join him in playing the cooperative strategy. Following this strategy in our illustrative example, you would make 6 while "learning," and 5 thereafter. There is some evidence that this is what actually happened in the Flood experiment.

An amusing way to exploit the inability of players to get together was suggested by Luce and Raiffa. They would have a company offer each of two players two alternative strategies: a "safe" strategy and a "double-cross" strategy. If both play safe, they each receive a dollar; if both play double-cross, they each lose a nickel; and if one plays safe while the other plays double-cross, the safe player gets a dollar and the double-cross player gets a thousand dollars. Luce and Raiffa felt that as long as the players were prevented from communicating and only played the game once, the company would get some free advertising at no great expense.

The Luce-Raiffa Advertising Model

Player *II*

		Safe	Double-Cross
Player *I*	Safe	($1,$1)	($1,$1,000)
	Double-Cross	($1,000,$1)	(−5¢,−5¢)

Figure 24

On the basis of the games we have seen up to now, the ability to communicate would seem to be an advantage. So far, the only messages contemplated are offers to cooperate, and these obviously must be in the interest of both players. Otherwise, the offers would not be made or, if made, would not be accepted. But it is also possible to communicate threats. The game shown in Figure 25 was devised by Luce and Raiffa. Compare what happens when the players can communicate and what happens when they can't.

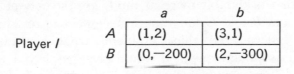

Figure 25

If the players can't communicate, they obviously can't threaten. Player *I* can do no better than to play strategy *A*, and player *II* can do no better than to play strategy *a*. But when the players are allowed to communicate, there is a radical change. Just what will happen is not altogether clear; to some extent, it will depend upon circumstances we haven't discussed. But, in any case, player *II*'s position deteriorates. If agreements made by the players can be enforced, player *I* can threaten to play strategy *B* unless player *II* commits himself to playing strategy *b*. If player *II* submits, player *I* will gain 2 and player *II* will lose 1, relative to the original (1, 2) payoff. If side payments are allowed, player *II*'s position becomes even worse. Not only can player *I* dictate player *II*'s strategy, but he can demand something under the table as well. True, player *II* can refuse to negotiate with player *I* and simply ignore his threats. It would still be in player *I*'s best interests to play strategy *A*, but who can say what would have a stronger influence on his behavior, his self-interest or his pique at being rebuffed? The point is that player *II* can avoid all these complications if communication is impossible.

It might be thought that communication would simplify the game, since the players would not have to guess at each other's intentions. But, in fact, games in which there is no communication, where the players must let their actions speak for them, are often less complicated than games in which bluffs, threats, and so on, can be passed on. An example of Schelling's shows why.

A picture is to be sold at an auction, and the bidding starts at $10. *B* says the picture is worth $15; *C* feels that it is worth considerably more. *B* and *C* are the only potential bidders, and *B* asks *C* for some inducement to refrain from bidding. Ignoring the ethical problems, what should *C* pay?

If there is no agreement, *B* would presumably bid as high as $15. If *B* refrains from bidding, *C* would get the picture for $10. So *B*'s agreement to stay out of the bidding would be worth $5 to *C*. On the other hand, *B* neither gains nor loses if he enters the bidding, so a "split-the-difference" approach would suggest that *C* pay *B* $2.50—that is, if *B*'s evaluation of the picture is honest. Actually, *C* doesn't know what the picture is worth to *B* or, for that matter, if it is worth anything to him at all. And even if *C* knows *B* doesn't want the picture, there is little he can do to prevent *B* from blackmailing him, short of letting *B* buy the picture when he enters the bidding—and this could only be done at the cost of giving up the picture himself. When the players can communicate, the game becomes one of bluff and counter-bluff.

But where there is no communication, there are no threats. *B* can stay out of the bidding, but he can't later demand a reward from *C*, since none was promised. Or, alternatively, he can be a spoiler and enter the bidding, but with no corresponding gain to himself. Thus, games without communication are considered by many to be more fundamental. And communication, which opens the way to the bluff and counterbluff, is regarded as an added complication.

The Order of Play

In the zero-sum game, the players pick their strategies simultaneously, neither player knowing his opponent's choice. If a player manages to find out his opponent's strategy in advance, he is far ahead and, in principle at least, the game is too trivial to be of interest. The non-zero-sum game is altogether different. The game may be far from trivial even when one player learns his opponent's strategy. What's more, the advantage of having this information may turn out to be a disadvantage. Let's look at an example.

A buyer and a seller are negotiating a contract in which the price per item and the quantity to be sold are still to be determined. According to established procedure, the seller first sets the price, which, once quoted, may not afterwards be changed; and the buyer indicates the quantity he wishes to buy.

In the present instance, a wholesaler can buy two items from the manufacturer, one for $4 and the other for $5. The retailer has two customers for these items, one of whom is willing to pay $9, and the other, $10. If the mechanism for negotiating is as we have just described it, what strategies should the players adopt? What will the outcome be?

Certain obvious characteristics of this game should be noted. It is clearly to both players' advantage to get together and in some way share the potential $10 profit. Also, if they are to share this profit equally and obtain a "fair" outcome, the selling price should be set at $7.

The wholesaler may have his eye on something better than a fair outcome, however. If he sets the price at $8 rather than $7, it would still be in the retailer's best interests to buy both items, although he would make a profit of only $3 rather than $5. (If he buys only one item, his profit would be $2, and if he buys nothing, he would have no profit at all.) In effect, the negotiating mechanism that requires the wholesaler to make the first move allows him to put pressure on the retailer, to his own advantage.

True, the retailer need not act mechanically in his own "self-interest" and allow himself to be exploited. Moreover, in the pure bargaining game in which buyer and seller negotiate freely about quantity and price simultaneously, the players will not always arrive at a selling price of $7; personality factors may affect the price one way or the other.

Even though the consequences of requiring the wholesaler to move first cannot be predicted precisely, however, the general effect is, clearly, to give the wholesaler the upper hand.

The Effect of Imperfect Information

The wholesaler-retailer game just described was one of a series of experiments studied in the laboratory. In the series, a variation of the game was also played. The original buying and selling prices were kept the same, but the rules were changed somewhat, and with them, the character of the game. In the new version, the wholesaler knew only his own profits; the retailer knew both players' profits. In addition, the retailer knew that the wholesaler did not know.

The basic difference between the variation and the original game was the reaction of the retailer when the wholesaler quoted a high selling price. In the original game, in which both players were fully informed, a high price on the wholesaler's part was interpreted by the retailer as a sign of greed, and as a result he often refused to go along. But when the retailer knew that the wholesaler was unaware of how large a share of the total profit he was asking, the retailer generally accepted his fate philosophically and did the best he could under the circumstances. Thus, the wholesaler often did better—and the retailer, worse—when he had less information.

A corollary of this is that it is often to a player's advantage to see that his partner is well informed. Suppose, in a labor-management dispute, that labor's demands are such that if granted, they would force the company out of business. In such an instance, the company should see to it that the union is informed of the effect of its demands. Of course, the company's interest is not so much to have the union know the truth for its own sake as it is to have the union believe that its goals are unattainable; and for this purpose, a lie will do just as well. The company, then, may try to deceive the union about, say, its inability to compete if it grants a raise. Or it may lie about its utility function, stating that it would prefer a prolonged strike to granting a raise, when in fact it would not. The union, for its part, might exaggerate the size of its strike fund. Ralph Cassady, Jr., describes some of the tactics used by competitors in a taxi-cab price war to "inform" (misinform) their opponents. They printed signs—which were never intended to be used —quoting rates considerably below the prevailing ones, and left them where they were sure to be seen by the competition. They also leaked the "information" that the company's owner had inherited a fortune—actually, he had inherited only a modest sum of money—and so gave the impression that the company had the capacity and the intention of fighting for some time.

In such a situation, a player can gain if he can convince his opponent that he has certain attitudes or capabilities, whether he really has them or not. (If you are bargaining for an antique that you particularly want, it is a good idea not to let the seller know it.) If he really has these capabilities or attitudes, we have the situation we spoke of earlier, where a player gains when his partner is better informed.

The Effect of Restricting Alternatives

A player is sometimes prevented from using some of his strategies. It is one of the paradoxes of non-zero-sum games that this restriction of a player's choice may be turned to his advantage. On the face of it, this seems absurd. How is it possible for a player to gain if he is prevented from using certain strategies? If it were to his advantage to avoid using certain strategies, couldn't he get the same effect simply by not using them? No, he could not; pretending you don't have certain alternatives is not the same as not having them. Suppose, for example, in the labor-management dispute, that there were strict wartime controls in effect that prevented the company from raising wages. In that case, a union that under different circumstances might go on strike would very likely continue working, without objection. By the same token, it would be to the woman's advantage, in our "battle of the sexes" example, if she tended to faint at the sight of blood and could not, therefore, sit through a prize fight.

When circumstances do not restrict a player's alternatives, he may try to gain an advantage by unilaterally restricting them himself. This is not always so effective, however. If the woman in the "battle of the sexes" commits herself and her husband to the ballet by buying two tickets in advance, her husband may refuse to accompany her out of pique. If she cannot attend a fight because she faints at the sight of blood, a factor over which she has no control, he may view it differently.

This principle of limiting your alternatives to strengthen your position can be applied in other ways. We have already seen one application in the "business partnership" example. Another application is the hypothetical weapon called the doomsday machine. A weapon of great destructive power, it is set to go off *automatically* whenever the nation that designed it is attacked.

The point of building a doomsday machine is this: so long as the defending nation preserves its option to withhold retaliation, the way is open to an unpunished attack. A potential aggressor may be tempted to attack and then, by threatening an even worse attack, inhibit the defending nation from retaliating. With the doomsday machine, the defending nation forecloses one of its options; it cannot help but retaliate. Here again, it is to a player's advantage to keep his partner well informed; if you have a doomsday machine, it is a good idea to let everyone know it.

Threats

A *threat* is a statement that you will act in a certain way under certain conditions. It is like the doomsday machine in that it limits your future actions: "If you cut your price by five cents, I'll cut mine by a dime." But it is different because this self-imposed restriction isn't binding; you can always change your mind. The purpose of a threat is to change someone's behavior: to make him do something he would not do otherwise. If the threat is carried out, it will presumably be to the detriment of the party that is threatened, but often it is also to the disadvantage of the party making the threat.

A threat is effective only to the extent that it is plausible. The greater the price the party making the threat must pay to carry it out, the less plausible the threat. This leads to the following paradox: if the penalty to the person making the threat is very high, he will be reluctant to burn his bridges behind him by committing himself irrevocably to the threat. But it is precisely the threatener's failure to burn his bridges which is the greatest inducement to ignoring his threat. When you are bargaining for a new car, you can discount such statements as "I won't let it go for a penny under $2,000." If the seller is convinced that at this price he will make no sale, he may later choose to ignore his own threat. In a store where prices are fixed by the management, however, and the final decision is made by a disinterested saleslady on salary, the "threat" of not lowering prices is essentially irrevocable. Of course, in that case, the store must accept the possibility that it will occasionally lose a sale. In the "battle of the sexes," one of the players may threaten to go it alone, but, again, the threat is not irrevocable, and the other player may threaten to resist. In the wholesaler-retailer example, the wholesaler was committed to his quoted selling price by the rules of the game and, as a result, the threat was much stronger.

Often, both players are in a position to threaten. In a bargaining game, for example, both the buyer and the seller can refuse to complete the sale unless the price is right. In the "battle of the sexes," each player could threaten to attend his or her preferred entertainment alone. Occasionally, however, it happens that only one of the players is in a position to threaten. Such a situation was cleverly exploited by Michael Maschler in his analysis of an inspection model designed to detect illicit nuclear testing.

The two players represented in this model are countries that have signed an agreement outlawing nuclear testing. One of them is considering violating the treaty; the other wants to be able to detect a violation if it occurs. (In reality, a country may play either role, or both at the same time.) The inspecting country has the benefit of detecting devices that indicate natural as well as artificial disturbances; it is allotted a quota of on-site inspections. The strictly mathematical problem consists of timing the inspections so as to maximize the probability of discovering a violation if there is one, and determining when, if ever, the potential treaty violator would test his device.

This, obviously, is not a zero-sum game, since presumably both the inspector and the inspected would prefer that there be no violations to having violations that are subsequently discovered. In any case, this is assumed in the model. Maschler showed that the inspector actually does best by announcing his strategy in advance and keeping to it, much as the wholesaler gained by setting a high price. (This is based on the assumption that the violator believes the announcement and then acts according to his own self-interest. There is no reason why the violator shouldn't believe it, for it is in the inspector's self-interest to tell the truth.) Why can't the tester use similar tactics; that is, announce his intention to cheat in some set manner and have the inspector make the best of it? If you look at the bare payoff matrix, it is clear that he can, but the political realities are such that this is not feasible.

Binding Agreements and Side Payments

When players negotiate, they often reach some sort of agreement. In some games there is no mechanism for enforcing agreements and the players can break their word with impunity. But in other games this is not the case. Though any agreement that is reached is reached voluntarily, once made, it is enforced by the rules. This possibility of making *binding agreements* has a strong influence on the character of the game.

Let us look at an anecdote related by Merrill M. Flood. Though this "game" actually had more than two players, it is pertinent here.

Flood wanted one of his children to baby-sit. He suggested that a fair way of selecting the sitter and setting a price for the service was to have the children bid against one another in a backwards auction. That is, he would start with a price of $4—the most he was willing to pay—and the children would bid successively, each bid lower than the last one, until the bidding stopped. The last person to bid would baby-sit for the agreed price.

It wasn't long before the children hit upon the possibility of collusion. When they asked about it, their father said he would allow them to "rig" the bidding if they satisfied two conditions: the final price must be no more than the ceiling of $4 he had set originally, and the children must agree among themselves in advance who would do the baby-sitting and how the money would be divided. As it happened, the children did not reach an agreement. Several days later, a bona fide auction was held and the final price was set at 90 cents. Thus, the mere opportunity to communicate and the existence of a mechanism to enforce agreements are not sufficient to guarantee that an agreement will be reached. In this game, the failure to get together resulted in an outcome markedly inferior to what might have been if the players had acted together.

In some games, it is possible for one player to affect the actions of another by offering him a "side payment," a payment made "under the table." This was the case in Flood's baby-sitting example. If an agreement had been reached, the child who baby-sat would have paid the other children a certain amount in return for staying out of the bidding.

In many games, however, the players cannot or may not make side payments. Sometimes this is a matter of policy—when the government invites private companies to bid on a contract, it responds to a cooperative strategy on the bidders' part (and to the corresponding side payments) with less than enthusiasm. Sometimes side payments are impractical because there is no unit that can be transferred from one player to the other. In the "battle of the sexes," the pleasure the wife feels when her husband takes her to the ballet is simply not transferable. (She may be able to reciprocate in a future game, however; perhaps next time she'll go to the fights.) Similarly, one legislator is barred from giving another any direct payment such as money in return for support, but he may repay him in kind in the future.

The simple example in Figure 26 will help clarify the role played by side payments. If it is impossible to make side payments, player *II* can do no better than choose strategy *B* and take a dollar. But if side payments are possible (and if the players can make binding agreements), it is a very different game. Player *II* is then in a position to demand a sizable portion of player *I*'s thousand dollars, and if player *I* refuses, he may be left with only $100. Whether player *II* will actually follow through on his threat and sacrifice a dollar, player *I* must decide for himself.

One cannot predict with any degree of assurance what will happen in this kind of game, or even to prescribe what should happen in theory. In practice, what happens no doubt depends strongly on the prize at stake. In games (*A*) and (*B*) in Figure 27, the players may communicate and may make binding agreements, but side payments are not permitted.

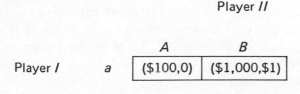

Player *II*

		A	B
Player *I*	a	($100,0)	($1,000,$1)

Figure 26

In game (*A*), player *II* will very likely try to get $1,000 by playing strategy *B* and persuading player *I* to play strategy *B*. Player *II*'s threat it to play strategy *A*, which will reduce player *I*'s payoff to nothing, at a cost of only $10 to

Game (*A*)

Player *II*

		A	B
Player *I*	a	(0,0)	($1,000,$10)
	b	(0,0)	($900,$1,000)

Game (*B*)

Player *II*

		C	D
Player *I*	c	(0,0)	($1,000, $800)
	d	(0,0)	($900, $850)

Figure 27

himself. Prudence seemingly dictates that player *I* should settle for the $900.

In game (*B*), the discussion might go much the same way, but now player *II*'s threat is less plausible. His "spite" strategy, strategy *C*, is a lot more expensive, though his threat to player *I* is essentially the same. Still, it is difficult to say what the players will do. If player *I* does decide to call player *II*'s bluff, however, he plays a more dangerous game in (*A*) than in (*B*).

An Application to a Prison Break

A vivid illustration of some of these factors at work in real life was detailed in a news article in *The New York Times* on June 28, 1965. The article described a prison disturbance in which two guards were taken hostage. The warden refused to negotiate with the prisoners as long as the guards were held captive, and the guards were eventually freed, unharmed. The warden was quoted as saying: "They wanted to make a deal. I refuse to make deals at any time. Therefore I didn't listen, and I don't know what the deal was about." Figure 28 analyzes the "game."

The Warden

		Free Prisoners	Do Not Free Prisoners
	Harm Guards	A	C
Prisoners	Do Not Harm Guards	B	D

Figure 28

We can start by ruling out *A*, since the prisoners gain nothing by harming the guards if they are released. Of the three remaining possibilities, the prisoners most prefer *B*, then *D*, and, least of all, *C*, which would mean additional punishment without any compensating gain. The warden wanted *D* the most, then *B* (presumably), and finally *C*.

The only chance of freedom the prisoners had was to threaten to harm the guards unless they were released, and to hope that the threat would be believed. But the warden simply cut communications; he refused to hear "what the deal was about." In effect, he committed himself to a strategy, not freeing the prisoners, and forced them to make a choice between *C* and *D*. His hope was that the prisoners would make the best of a bad situation by choosing *D* rather than *C*. And so they did—but if they had been sufficiently vindictive, they might have acted otherwise. Whether the warden made the proper choice is a question about which reasonable men may differ. Indeed, the warden and the hostages probably did.

The Prisoner's Dilemma

Two men suspected of committing a crime together are arrested and placed in separate cells by the police. Each suspect may either confess or remain silent, and each one knows the possible consequences of his action. These are: (1) If one suspect confesses and his partner does not, the one who confessed turns state's evidence and goes free and the other one goes to jail for twenty years. (2) If both suspects confess, they both go to jail for five years. (3) If both suspects remain silent, they both go to jail for a year for carrying concealed weapons—a lesser charge. We will suppose that there is no "honor among thieves" and each suspect's sole concern is his own self-interest. Under these conditions, what should the criminals do? The game is shown in Figure 29. This is the celebrated prisoner's dilemma which was originally formulated by A. W. Tucker and which has become one of the classical problems in the short history of game theory.

Suspect *II*

		Confess	Do Not Confess
	Confess	(5 yrs, 5 yrs.)	(go free, 20 yrs.)
Suspect *I*	Do Not Confess	(20 yrs., go free)	(1 yr., 1 yr.)

Figure 29

Let us look at the prisoner's dilemma from the point of view of one of the suspects. Since he must make his decision without knowing what his partner will do, he must consider each of his partner's alternatives and anticipate the effect of each of them on himself.

Suppose his partner confesses; our man must either remain silent and go to jail for twenty years, or confess and go to jail for five. Or, if his partner remains silent, he can serve a year by being silent also, or win his freedom by confessing. Seemingly, in either case, he is *better off confessing!* What, then, is the problem?

The paradox lies in this. Two naïve prisoners, too ignorant to follow this compelling argument, are both silent and go to prison for only a year. Two sophisticated prisoners, primed with the very best game-theory advice, confess and are given five years in prison in which to contemplate their cleverness.

We will return to this argument in a moment, but before we do, let's consider the essential elements that characterize this game. Each player has two basic choices: he can act "cooperatively" or "uncooperatively." When all the players act cooperatively, each does better than when all of them act uncooperatively. For any fixed strategy(ies) of the other player(s), a player always does better by playing uncooperatively than by playing cooperatively.

In the following examples, taken from many different contexts, these same basic elements appear.

1. Two different firms sell the same product in a certain market. Neither the product's selling price nor the total combined sales of both companies vary from year to year. What does vary is the portion of the market that each firm captures, and this depends on the size of their respective advertising budgets. For the sake of simplicity, suppose each firm has only two choices: spending $6 million or $10 million. The size of the advertising budget determines the share of the market and, ultimately, the profits of each company, as follows:

If both companies spend $6 million, they each get a $5 million profit. If a company spends $10 million when its competitor spends only $6 million, its profit goes up to $8 million at the expense of its competitor, who now loses $2 million. And if both companies spend $10 million, the extra marketing effort is wasted, since the market is fixed and the relative market position of each company remains the same; consequently, the profit of each company drops to $1 million. No collusion is allowed between the firms. The game, in normal form, is shown in Figure 30.

Corporation *II*

		Spend $6 Million	Spend $10 Million
Corporation *I*	Spend $6 Million	($5 million, $5 million)	(−$2 million, $8 million)
	Spend $10 Million	($8 million, −$2 million)	($1 million, $1 million)

Figure 30

2. There is a water shortage and citizens are urged to cut down on water consumption. If each citizen responds to these requests by considering his own self-interest, no one will conserve water. Obviously, any saving by an individual has only a negligible effect on the city's water supply, yet the inconvenience involved is very real. On the other hand, if everyone acts in his own self-interest, the results will be catastrophic for everyone.

3. If no one paid his taxes, the machinery of government would break down. Presumably, each person would prefer that everyone pay his taxes, including himself, to having no one pay taxes. Better yet, of course, everyone would pay taxes except the individual himself.

4. After several years of overproduction, farmers agree to limit their output voluntarily in order to keep the prices up. But no one farmer produces enough seriously to affect the price, so each starts producing what he can and selling it for what it will bring, and once again there is overproduction.

5. Two unfriendly nations are preparing their military budgets. Each nation wants to obtain a military advantage over the other by building a more powerful army, and each spends accordingly. They wind up having the same relative strength, and a good deal poorer.

As we can see from these examples, this kind of problem comes up all the time. For convenience sake, let us fix our attention on a single game, using the first example, that of two companies setting their advertising budgets. The game will be the same, except for an additional condition. The budget will not be set once but, more realistically, we will assume that it is fixed annually for a certain period of time; say, twenty years. When each company decides on its budgets for any given year, it knows what its competitor spent in the past.

In discussing the "prisoner's dilemma," we had decided that if the game is played only once, the prisoners have no choice but to confess. The same line of reasoning applied here leads to the conclusion that the firms should each spend $10 million. But when the game is played repeatedly, the argument loses some of its force. It is still true that if you spend $10 million in a given year, you will always do better than if you spend $6 million *that year*. But if you spend $10 million in one year, you are very likely to induce your competitor to spend $10 million the next year, and that is something you don't want. A more optimistic strategy is to signal your intent to cooperate by spending $6 million and hope your competitor draws the proper inference and does the same. This strategy could lead to a cooperative outcome and in practice often does. But in theory there is a problem.

The argument that spending $6 million in any one year tends to encourage your competitor to spend the same amount the next year is all very well for the first nineteen years, but it clearly breaks down in the twentieth year. In the twentieth year, there is *no next year*. When the firms reach the twentieth year, they are in effect in the same position they are in when they play the game only once. If the firms want to maximize their profits, and we assume that they do, the argument favoring the uncooperative strategy is as compelling now as it was then.

But the argument doesn't end there. Once the futility of cooperating in the twentieth year is recognized, it follows that there is no point in cooperating in the nineteenth year either. And if there is no chance of inducing a cooperative response in the nineteenth year, why cooperate in the eighteenth year? Once you fall into the trap, you fall all the way: there is no point in cooperating in the eighteenth, the seventeenth . . . or the first year, either. If you accept the argument supporting the uncooperative strategy in the single instance, it follows that you must play uncooperatively not only in the last of a series but in every individual trial as well.

It is when the "prisoner's dilemma" is played repeatedly —and not for a fixed number of trials but for an indefinite period—that the cooperative strategy comes into its own. And these are precisely the conditions under which the "prisoner's dilemma" is often played. Two competing firms know that they won't be in business forever, but they generally have no way of knowing when death, merger, bankruptcy, or some other force will end their competition. Thus, the players can't analyze what will happen in the last "play" and then work backward from there, for nobody knows when the last "play" will be. The compelling argument in favor of the uncooperative strategy breaks down, then, and we breathe a sigh of relief.

This is really the point. The "prisoner's dilemma" has one characteristic that makes it different from the other games we have discussed. As a rule, when analyzing a game, one is content if one can say what rational players should do and predict what the outcome will be. But in the "prisoner's dilemma" the uncooperative strategy is so unpalatable that the question most people try to answer is not: What strategy should a rational person choose? but: How can we justify playing a cooperative strategy? Many different answers to this last question have been proposed. Let us consider a few.

The "Prisoner's Dilemma" in the Past

The "prisoner's dilemma" is important because it presents in capsule form a problem that arises in many different ways. Some of the manifestations of the "prisoner's dilemma" were discussed long before there was a theory of games. Thomas Hobbes, the political philosopher, examined one version of the dilemma in which the "players" were the members of society.

Society, Hobbes conjectured, was originally in a state of anarchy. Constant warfare and banditry were the consequence of each individual's trying to further his own narrow self-interest; it was a society in which B might murder C for a shiny bauble, and B might be murdered in turn for the same reason. It would be to everyone's individual advantage, Hobbes felt, if restrictions were imposed and enforced; that is, B would prefer to give up the chance of getting the bauble in return for additional security. Hobbes saw the social contract as an enforced cooperative outcome. In *The Leviathan* he described the creation of a government (preferably a monarchy) ". . . as if every man should say to every man, I authorize and give up my right of governing myself, to this man or to this assembly of men, on this condition that thou give up thy right to him, and authorize all his actions in a like manner." (Whether historically this is an accurate picture is unimportant; what matters is how the problem was perceived and how it was resolved.)

Hobbes, after pointing out the disadvantages of the unco-operative outcome, suggests that the many independent decisions to cooperate or not be taken out of the hands of the people that make up society. In effect, society should submit to compulsory arbitration, and the government should play the role of the arbiter. This is not an uncommon point of view. Luce and Raiffa, in *Games and Decisions,* make the same point: "Some hold the view that one essential rule of government is to declare that the rules of social 'games' be changed whenever it is inherent in the game situation that the players, in pursuing their own ends, will be forced into a socially undesirable position."

On a more modest scale, the noted sociologist George Simmel recognized that competing businesses are often faced with what amounts to a "prisoner's dilemma." He described the behavior of businessmen who, in effect, are playing this game for an indefinite number of times:

"Inter-individual restriction of competitive means occurs when a number of competitors voluntarily agree to renounce certain practices of outdoing one another—whereby the renunciation by one of them is valid only so long as the other too observes it. Examples are the arrangement among book-sellers of a given community to extend none or more than five or ten percent discount; or the agreement among store owners to close their businesses at eight or nine o'clock, etc. It is evident that here mere egoistic utility is decisive: the one does without those means of gaining customers because he knows that the other would at once imitate him if he did not, and that the greater gain they would have to share would not add up to the greater expenses they would likewise have to share. . . . In Economics, the third party is the consumer; and thus it is clear how the road toward cartelization is taken. Once it is understood that one can do without many competitive practices provided the competitor does likewise, the result may not only be an even more intense and purer competition, which has already been emphasized, but also the opposite. The agreement may be pushed to the point of abolishing competition itself and of organizing enterprises which no longer fight for the market but supply it according to a common plan. . . . This teleology, as it were, transcends the parties, allowing each of them to find its advantage and achieves the seeming paradox that each of them makes the opponent's advantage its own."

Note that Simmel looks at the consumer as an outsider, who, though he is affected by what the competing businesses do, has no control over what happens. In effect, the consumer is not a player. While cooperation between businesses is easily turned to their mutual advantage, the effects can easily be anti-social as far as society as a whole is concerned. And so society prohibits "cooperative play" in the form of trusts, cartels, price fixing, and bribery.

This theme—the tendency of competitors to avoid mutually destructive price competition—is also considered by John Kenneth Galbraith. In *American Capitalism: The Concept of Countervailing Power*, he says: "The convention against price competition is inevitable. . . . The alternative is self-destruction." Price competition, then, tends to be replaced by competition in sales and advertising. When there are only a few sellers, inter-industrial competition diminishes. If a few large industries are bargaining with their workers on wages, there ceases to be a wage competition among the individual industries to attract workers. There is, rather, bargaining between the "countervailing powers" of labor and management.

Problems of the "prisoner's dilemma" type, in one form or another, have been around for some time. The cooperative strategy is generally accepted as the "proper" one (except when its effect is anti-social), sometimes for ethical reasons. Immanuel Kant asserted that a person should decide if his act is moral by examining the effect of everyone's acting similarly; the golden rule says much the same thing. More recently, Professor Rapoport in *Fights, Games and Debates* asserted that there are considerations other than a player's narrow self-interest which he should take into account when he chooses his play. If there is to be any hope of reaching the elusive cooperative outcome, Rapoport feels, it is necessary that the players accept certain social values, and having accepted these values, players should cooperate even in the one-trial version of the "prisoner's dilemma." His argument goes like this:

"Each player presumably examines the whole payoff matrix. The first question he asks is 'When are we both best off?' The answer in our case is unique: at [the cooperative outcome]. Next 'What is necessary to come to this choice?' Answer: the assumption that whatever I do, the other will, made by both parties. The conclusion is, 'I am one of the parties; therefore I will make this assumption.'"

Rapoport is well aware that his point of view is in conflict with "'rational' strategic principles" of self-interest; he simply rejects these principles. The minimax strategy in the two-person, zero-sum game, he asserts, is also based on an assumption: that one's opponent will act rationally; that is, in accordance with his own self-interest. If in this game one's opponent fails to act rationally, the minimax strategy will fail to exploit his errors. Just as you may be mistaken when you assume your opponent will be rational in the zero-sum game, you may be mistaken in the non-zero-sum game when you anticipate your partner's good will.

Since most people are generally reluctant to accept the uncooperative strategy as the proper one, there is a temptation, whenever a way out is proposed, to take it. Despite my sympathy with Rapoport's effort, however, I do not believe the paradox of the "prisoner's dilemma" has really been overcome. To see why, let us go back a bit.

In the "prisoner's dilemma" examples discussed in this

chapter, the payoffs were stated in terms of "years in prison" or net profits, rather than utiles. The actual utilities involved are only suggested by such statements as: "Each player is only concerned with his own self-interest," or "Each company only wants to maximize its own profits." For the sake of simplicity, a more formal description in terms of utilities was avoided. But at bottom are certain assumptions that, though stated only inexactly, are critical. If they are not valid, we may be playing a radically different game from the one we think we are playing. If, in the original prisoner's dilemma, a prisoner would prefer to spend a year in jail along with his partner rather than go free, knowing his partner was serving twenty years, the argument for confessing breaks down. But then the game could hardly be called a prisoner's dilemma. The original paradox has in fact not been resolved. This is the objection to Rapoport's argument: his Golden Rule approach assumes the problem out of existence. If a player is as concerned with his partner's payoff as he is with his own, the game isn't a prisoner's dilemma; and if each player is interested solely in his own payoff, Rapoport's comments aren't pertinent.

The analogy between the assumptions that are made in the zero-sum and in the non-zero-sum games is not very convincing either. In the zero-sum game, you can get the value of the game whether your opponent is good, bad, or indifferent; you do not have to assume that he is rational. Rapoport states as much but goes on to say that by playing minimax you lose the opportunity of exploiting errors, and with a foolish opponent you should not be content to get only the value of the game.

This is not entirely true. There are games—we have seen some of them already—in which an inferior strategy on one player's part and the minimax strategy on the other's will lead to a payoff which is greater than the value of the game for the minimax player. But, even in those games in which playing minimax precludes your getting more than the value of the game, the analogy is questionable. In order to exploit your opponent's weaknesses, it is not enough to know he will deviate from the minimax; you must also know how. Suppose, for example, you are going to match pennies *once* with a simpleton and you decide to assume that he will *not* play rationally. Specifically, you believe that he is quite capable of playing one side of the coin with a probability greater than one-half. But which side? And how can you exploit his inferior strategy if you don't know which side?

As a rule, when you play a minimax strategy in a zero-sum game it is not because you have faith in your opponent's rationality but because you have no other, more attractive alternative—and this is true even when you suspect your opponent is capable of making a mistake.

In the "prisoner's dilemma," however, the assumption that your partner will cooperate is really an assumption. Unless you are a masochist, if you play cooperatively you must believe as an act of faith that your partner will too. Even if your partner cooperates and seemingly justifies your act of faith, some players will still question your choice since you could have done better still by playing uncooperatively. This attitude may seem piggish, but, then, players don't look to game-theorists for moral principles; they already have their own. All they ask is to find a strategy that will suit their purpose, selfish or otherwise.

The difference between the assumptions made in the zero-sum and in the non-zero-sum games is even more clear when they fail. And there is no question that people often *do* fail to play cooperatively in society's "prisoner's dilemmas." In non-zero-sum games, cooperating with a partner who doesn't cooperate with you leads to disaster; in zero-sum games, the worst that can happen when you play minimax is that you lose an opportunity to swindle your opponent.

The Nash Arbitration Scheme

A player in a bargaining game is in an awkward position. He wants to make the most favorable agreement that he can, while avoiding the risk of making no agreement at all; and, to a certain extent, these goals are contradictory. If one party indicates a willingness to settle for any terms, even if his gain is only marginal, he will very likely arrive at an agreement, but not a very attractive one. On the other hand, if he takes a hard position and sticks to it, he is likely to reach a favorable agreement if he reaches any agreement at all—but he stands a good chance of being left out in the cold. A car dealer who is anxious to sell a car will hide this fact from a buyer; but he will go to great lengths to determine how much he must drop his price to make a sale—even to the extent of using hidden microphones.

Even when a player is reconciled to receiving only a modest gain and pushes hard for an agreement, his eagerness is often interpreted as weakness, leading his partner to stiffen his demands and actually lessening the chance of an agreement. This is what often happens when one of two warring nations sues for peace. In a labor-management dispute that actually took place, one party who had a weak heart offered favorable terms in order to reach an immediate settlement. What happened was very different from what had been anticipated. Instead of waiving the usual bargaining procedure, the other party became suspicious and then resistant. The terms of the ultimate settlement were identical to those originally offered, but they were arrived at only after very hard bargaining.

One way of circumventing the actual bargaining process, at least in principle, is to have the terms of the agreement set by arbitration. In this way, you can avoid the danger of not reaching a settlement at all. The problem is to establish arbitration that somehow reflects the strengths of the players in a realistic way, so that you get the effects of negotiation without the risk. John Nash suggests the following procedure.

He begins by assuming that two parties are in the process of negotiating a contract. They might be management and labor, two countries formulating a trade agreement, a buyer and seller, etc. For convenience, and with no loss of generality, he assumes that a failure to agree—no trade, no sale, a strike, etc.—would have a utility of zero to both players. Nash then selects a single arbitrated outcome from all the agreements that the players have it in their power to make: that outcome in which the product of the players' utilities is maximized.* This scheme has four desirable properties which he feels justify its use, and it is the *only* one that does. The four properties are:

1. *The arbitrated outcome should be independent of the utility function.* Any arbitrated outcome should clearly depend on the preferences of the players, and these preferences are expressed by a utility function. But, as we saw earlier, there are many utility functions to choose from. Since the choice of utility function is completely arbitrary, it is reasonable to demand that the arbitrated outcome not depend on the utility function selected.

2. *The arbitrated outcome should be Pareto optimal.* Nash considered it desirable that the arbitrated outcome be Pareto optimal; that is, that there should not be any other outcome in which both players simultaneously do better.

3. *The arbitrated outcome should be independent of irrelevant alternatives.* Suppose there are two games A and B in which every outcome of A is also an outcome of B. If the arbitrated outcome of B turns out also to be an outcome

---

* It should be noted that players may not only obtain utility pairs associated with simple agreements but may also obtain intermediate pairs by using coordinated strategies as well. Suppose, for example, that in the "battle of the sexes," going to the ballet has a utility of 4 for the man and 8 for his wife, and going to the fights has a utility of 6 for the man and 2 for his wife. Each may obtain a utility of 5 if they let a toss of a coin determine the choice of the evening's entertainment.

of A, this outcome must also be the arbitrated outcome of A. Put another way, the arbitrated outcome in a game remains the arbitrated outcome even when other outcomes are eliminated as possible agreements.

4. *In a symmetric game, the arbitrated outcome has the same utility for both players.* Suppose the players in the bargaining game have symmetric roles. That is, if there is an outcome which has a utility of x for one player and a utility of y for the other, there must also exist an outcome which has a utility of y for the first player and x for the second player. In such a game, the arbitrated outcome should have the same utility for both players.

Before the Nash arbitration scheme can be applied, the utility function of both players must be known. This is its biggest disadvantage, for not only are the utilities not always known; they are often deliberately obscured by the players. If a player's utility function is misrepresented, it can be turned to his advantage. This is reassuring in a way, for it suggests the scheme is realistic. In real life also, the utility function is often misrepresented, as we have seen.

It is important to realize that the Nash scheme is neither enforceable nor a prediction of what will happen. It is, rather, an a priori agreement obtained by abstracting away many relevant factors such as the bargaining strengths of the players, cultural norms, and so on. (In this respect, it is similar to the Shapley value that we shall discuss later.) As a matter of fact, the Nash outcome often appears to be unfair: it tends to make the poor poorer, and the rich richer. This is to be expected, however. A rich player is often in a stronger position than a poor one. To see how this works, consider the following example.

Suppose a rich man and a poor man can get a million dollars if they can agree on how to share it between them; if they fail to agree, they get nothing. In such a case, the Nash arbitration scheme would generally give the rich man a larger portion than it would give the poor man, because of the difference in their utility functions. Let's take a moment to see why.

When *relatively* large amounts of money are involved—that is, amounts of money that are large relative to what a person already possesses—people tend to play safe. Most people, unless they are very rich, would prefer a sure million dollars to an even chance of getting $10 million, although they would prefer an even chance of getting $10 to a sure dollar. But a large insurance company would prefer the even chance of $10 million, and in fact gladly accepts much less attractive risks every day. This indifference to the difference between large sums of money is reflected in the poor man's utility function much more strongly than in the rich man's. The relative attractiveness of $1 and $10 to the poor man would be like the relative attractiveness of $1 million and $10 million to a man who is very wealthy. A utility function which correctly reflects the poor man's situation would be the square-root function: $100 would be ten utiles, $1 would be one utile, $16 would be four utiles, and so on. Thus, the poor man would be indifferent between an even chance at $10,000 and a sure $2,500. (The specific choice of the square-root function is of course arbitrary; many others would do as well.) It may be assumed that the rich man's utility function is identical to the money in dollars. Under these conditions, Nash's suggested outcome would be that the rich man gets two-thirds of the million dollars and the poor man only a third.

Two-Person, Non-Zero-Sum Game Experiments

One reason for studying experimental games is that they are interesting. When you spend a lot of time thinking about how people should behave in theory, you become curious about how they actually behave in practice. A second reason for studying experimental games is that you may gain insights that will enable you to play better. That is a much more important consideration in non-zero-sum games than it is in zero-sum games. In the two-person, zero-sum game, a player can obtain the value of the game by his own efforts alone; he doesn't have to worry about what his opponent does. In the non-zero-sum game, unless you are willing to be satisfied with a minimal return—as a buyer and a seller may have to be when they fail to reach an agreement—you *must* be concerned with how your partner plays. Similarly, in a sequence of "prisoner's dilemmas," what you anticipate will be your partner's play will affect your own.

Granted that it is worthwhile to learn more about how people behave, why should we turn to the laboratory rather than to actual life for our data? Examples of non-zero-sum games in everyday life are certainly common enough. Lawrence E. Fouraker and Sidney Siegel, in *Bargaining and Group Decision Making,* answer the question thus: "In the specific case of Bilateral Monopoly, it would be extremely unlikely that appropriate naturalistic data could be collected to test the theoretical models. This is not because the phenomenon is unusually rare. Indeed, there are numerous daily exchanges that are conducted under conditions that approximate Bilateral Monopoly; a franchised dealer negotiates with a manufacturer regarding quotas and wholesale price; two public utilities bargain about the division of some price they have placed on a joint service; a chain grocery store negotiates with a canner, who in turn must deal with farmers' cooperatives; labor leaders in a unionized industry deal with management in that industry; and so forth."

The trouble with "real" games is that they are not set up for our convenience. The variables are not controlled; we are not likely to find two situations that are identical except for one variable. This makes it difficult to determine just how important a variable is in influencing the final outcome. Also, it is usually not feasible to determine the payoffs. In the laboratory, on the other hand, the players can be separated to avoid personal contact (an unnecessary complicating factor), the payoffs are clear, and the variables can be altered at will. It is also possible to motivate the players by making the payoffs sufficiently large—in principle, at least.

Some Experiments on the "Prisoner's Dilemma"

The many experiments conducted on the "prisoner's dilemma" all share a common purpose: to determine under what conditions players cooperate. Among the significant variables that determine how a player will behave are the size of the payoffs, the way the other person plays, the ability to communicate, and the personality of the players. In a series of experiments by Alvin Scodel, J. Sayer Minas, David Marlowe, Harvey Rawson, Philburn Ratoosh, and Milton Lipetz which were described in three articles in the *Journal of Conflict Resolution* between 1959 and 1962, a "prisoner's dilemma" game and some variations on it were played repeatedly. Let us consider some of the observations of the experimenters.

The game shown in Figure 31, which we will call game *I*, was played fifty times by each of twenty-two pairs of players. *C* and *NC* are the cooperative and noncooperative strategies, respectively. The players were physically separated throughout the fifty trials, so no direct communication was possible. At every trial, each player knew what his partner had done on every previous trial.

At each trial, each player had two choices, making a total of four possible outcomes. If the players had picked their strategies at random, we would have expected the cooperative payoff (3, 3) 25 percent of the time; the uncooperative payoff (1, 1) 25 percent of the time; and one of the mixed payoffs (5, 0) or (0, 5), 50 percent of the time. In fact, the uncooperative outcome predominated. Of the twenty-two pairs, twenty had more uncooperative payoffs than they had any other combination. Even more surprising, the tendency of the players was to be more uncooperative as the game progressed.

Game *I*

	C	NC
C	(3, 3)	(0, 5)
NC	(5, 0)	(1, 1)

Figure 31

Game *Ia* was a repetition of game *I*, with one variation: the players were allowed to communicate in the last twenty-five of the fifty trials. As you would expect, the results on the first twenty-five trials were pretty much the same as before. On the last twenty-five trials, there was still a tendency not to cooperate, but it was not so pronounced as when the players couldn't communicate.

Game *II* had the same payoff matrix as games *I* and *Ia*, but a variation was introduced. The subjects didn't play against each other but against the experimenter, though they weren't aware of it. The experimenter played in accordance with a predetermined formula: at each trial, he did what the subject did. If a subject cooperated, the experimenter did also (on the same trial), and the subject received 3. If the subject didn't cooperate, he received only 1. The game was repeated fifty times, just as games *I* and *Ia* were. The players chose the uncooperative strategy 60 percent of the time, and played noncooperatively in the second twenty-five trials more often than they did in the first twenty-five trials.

One would think that in fifty trials the subjects would realize that they were not playing against just another person. And, presumably, subjects who caught on would play strategically. Both post-experimental interviews and the experimental results indicated, however, that every subject believed the responses of his "partner" to be legitimate. Those players who noticed a similarity between their own and their "partner's" play attributed it to coincidence.

In several other "prisoner's dilemma" games, the same pattern was repeated. In game *III* (see Figure 32), for example, a player only gained 2 by defecting from the cooperative outcome. Nevertheless, 50 percent of the time, in the first half of a thirty-trial sequence, the players were

Game *III*

	C	NC
C	*(8, 8)*	(1, 10)
NC	(10, 1)	(2, 2)

Figure 32

uncooperative. And in the last fifteen trials, the percentage went up to 65 percent. When the experiment was repeated with the second player replaced by the experimenter (who always played uncooperatively), the frequency of cooperative play was virtually unchanged. In game *IV* (see Figure 33), a "prisoner's dilemma" in which the payoffs were mostly negative and the players had to scramble to cut their losses, there was virtually no cooperation at all.

Game *IV*

	C	NC
C	(−1, −1)	(−5, 0)
NC	(0, −5)	(−3, −3)

Figure 33

Some of the most interesting experiments were not really "prisoner's dilemma" games at all. Three of these games are shown in Figure 34. Each of the three games was played in thirty-trial sequences, and in each game it was surprising how frequently the players failed to cooperate. In game V, the players failed to cooperate 6.38 times on the first fifteen trials (on average), and 7.62 times on the last fifteen trials: a small but statistically significant increase.

Game *V*

	C	NC
C	(6, 6)	(4, 7)
NC	(7, 4)	(−3, −3)

Game *VI*

	C	NC
C	(3, 3)	(1, 3)
NC	(3, 1)	(0, 0)

Game *VII*

	C	NC
C	(4, 4)	(1, 3)
NC	(3, 1)	(0, 0)

Figure 34

Games *VI* and *VII* turned out about the same. In game *VI*, the players were noncooperative slightly more than half the time and were a little more cooperative during the first half of the trials than they were in the second half. In game *VII*, the players cooperated about 53 percent of the time, but in the last fifteen trials *they failed to cooperate more than half the time.*

Throughout these experiments, there was a marked consistent tendency to play uncooperatively. Uncooperative play is understandable in games of the "prisoner's dilemma" type, where it has obvious advantages, at least in the short run. But the tendency persisted into the last three games, and this is much more difficult to explain.

In game V, uncooperative play was rewarded only part of the time: a player gained only if his partner cooperated —and not very much at that. And if his partner played uncooperatively also, he received the smallest possible payoff. In games VI and VII, it was absurd to play uncooperatively. In game VI, there was no chance of gaining, and some chance of losing, if you played uncooperatively, and in game VII, *a player who failed to cooperate always received a smaller payoff, whatever his partner did.* His partner's play affected only the amount that he lost. Despite this, in every game but the last, uncooperative play predominated. Even in the last game, the outcome was very close to what you would expect if the players had picked their strategies by tossing a coin. Moreover, as the game progressed, the tendency to cooperate became weaker rather than stronger.

Why players fail to cooperate is not altogether clear. A player might want to exploit his partner or might fear that his partner is about to exploit him. A player might not understand what the game is all about or might doubt that his partner does—though this last possibility doesn't seem very likely. If a player with insight into the game fails to cooperate because he fears his partner won't "read his message," he would presumably cooperate when he was permitted to talk to his partner and make his message plain. But, in fact, there was only a slight increase in cooperation when the players were allowed to communicate.

Not only were the players slow to cooperate with each other; they appeared to be almost oblivious to what their partners did. They were not even suspicious when their partner's play was identical to their own. Suspicious or not, their playing uncooperatively 60 percent of the time for

fifty-trial periods in the face of this behavior borders on
the incredible. *There was roughly the same amount of noncooperative play when the experimenter duplicated the subject's play as there was when the experimenter always played
noncooperatively.*

The subjects seemed to regard these games as purely
competitive: beating one's partner was most important, and
the player's own payoff was only secondary. This tendency
to compete, which has been observed by many experimenters
and which increased as the game progressed, has been attributed to boredom and the smallness of the monetary payoffs. One wonders if this determination to beat one's partner
would diminish if an appreciable amount of money were
at stake.

A Bargaining Experiment

Some time ago, Lawrence E. Fouraker and Sidney Siegel
ran a carefully controlled series of experiments on bargaining games, and their observations are included in their book
Bargaining Behavior. Because bargaining games often occur
in life (labor-management negotiations, trade agreements between nations, the purchase of a used car—all fall in this
category), and because the results of the experiments are of
such interest and value, we will study this work in some
detail.

The players in the bargaining game were a retailer and
a wholesaler. The wholesaler was in a position to buy twenty
items: five at $6, ten at $8, and another five at $10. The
retailer had twenty potential customers: four were willing
to pay $11 for an item, eight would pay $12, and another
eight were willing to pay $13. The bargaining always took
the following form: The wholesaler started by quoting a
price. This price remained fixed; once quoted, it could not
be withdrawn even with the wholesaler's consent. The retailer then decided how many items he wished to purchase,

if any, at this price. This decision ended the transaction, and each player kept whatever profit he had made. The players were not allowed to see each other or send messages.

All the games had this basic format, but there were many variations, which we will discuss later.

The "Optimal" Strategies

Before considering the results of the experiments, let us take a moment to examine the structure of the game. The greatest *joint* profit occurs when the wholesaler sells twenty items to the retailer. Since the wholesaler can buy the twenty items at $160, and the retailer can sell them for $244, the potential joint profit is $84.

Now let us see what happens if each player acts to maximize his own immediate profit and anticipates that his partner will do the same. The retailer has a simpler decision, since he moves last. If the retailer is interested only in his own profits, he should buy twenty items when the wholesaler sets the price below $11, sixteen items when the price is between $11 and $12, eight items when the price is between $12 and $13, and nothing when the price is over $13. If, for example, the wholesaler set the price at $11.50, the retailer would buy sixteen items, sell eight of them at $12 and eight of them at $13, and make a profit of $16. The wholesaler would have bought five of the items at a price of $6, ten at $8, and one at $10, at a total cost of $120, so his profit would be $64.

Assuming that the retailer plays his best strategy, the wholesaler's optimum strategy is also a matter of simple computation. He should set his price a hair under $12; the retailer will buy sixteen items and make a profit of just over $8; and the wholesaler's profit will be maximized: he will make $72. The relationships between the set price and the profits of the player are illustrated in Figure 35.

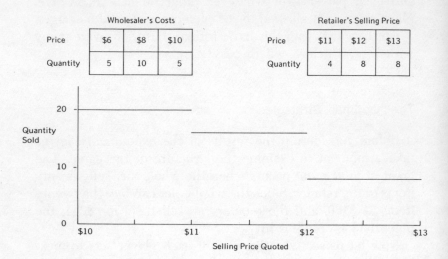

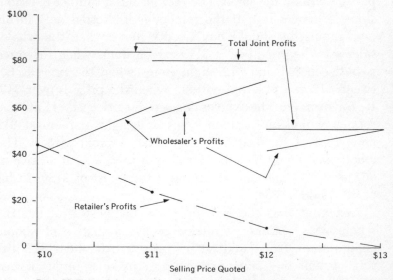

Figure 35. These graphs are based on the assumption that the retailer will choose the quantity that will maximize his profit.

Some Experimental Variables

The purpose of these experiments was to see what effect certain variables had on the players' behavior. The basic structure just described was used throughout the series, but the money payments were changed so that the influence of each variable could be observed.

When a large piece of land is divided into two parts (because of a treaty between nations, or in accordance with the terms of a will), certain kinds of borders are usually preferred. Borders usually consist of one or more straight-line segments or reflect natural landmarks such as rivers or lakes. In a similar way, negotiators often use "landmarks" when fashioning an agreement. These might be based on cultural norms or might reflect the joint welfare of the players.

One of the natural places to form an agreement is at the *equal-profits* point: that price at which each player makes the same profit. Although this price may not reflect the strengths of the players accurately, it would generally be accepted as a fair outcome. For the particular money amounts that we used, the equal-profit point is reached when the price is $10.10; at this price, all twenty items would be bought by the retailer and each player would obtain a profit of $42.

A second natural price would be the *Bowley point*. The Bowley point is defined in the following way. Suppose the wholesaler assumes that the retailer will always maximize his own profit once a price is quoted. On the basis of this assumption, the wholesaler calculates his own profit for every possible price and picks the best one from his point of view. This is the Bowley point. In the game we just described, the Bowley is obtained when the price is set at just under $12. At this price, sixteen items are sold, and the profits of the wholesaler and the retailer are $72 and $18, respectively.

Another desirable characteristic is *Pareto optimality*. An outcome is Pareto optimal if there is no other possible agreement that enables both players to do better *simultaneously*. Pareto optimal outcomes are not unique. In the game described, every outcome in which the retailer buys twenty items from the wholesaler is Pareto optimal. Conversely, these are the only outcomes that are Pareto optimal. When twenty items are sold, the total profit of the two players is $84. Since no larger total profit is possible, one player can do better only if the other does worse. On the other hand, if less than twenty items are sold, both players may do better by selling one more item and splitting the profit.

In general, the Bowley point need not be Pareto optimal. In our example, it isn't. If instead of a selling price of $12 (the Bowley point), at which sixteen items are sold, the selling price is set at $11.70 and twenty items are sold, the retailer's profit would go up from $8 to $10 and the wholesaler's from $72 to $74. (If the wholesaler actually quoted a price of $11.70, however, the retailer would do better to buy only sixteen items. His profit would then rise to $12.80 and the wholesaler's profit would drop to $67.20. This is why the Bowley point is a plausible outcome, even though it is not Pareto optimal.)

Fouraker and Siegel's concern, in running several variations of the basic experiment described earlier, was to determine the conditions under which the outcome would be the Bowley point rather than a Pareto optimal outcome.

One variable—the one that was most significant, as it happened—was the frequency with which the game was played. At times the game was played only once, and at other times it was played repeatedly. When the game was played only once and the wholesaler tried to move to the Bowley point at the retailer's expense, the retailer had these alternatives: he could make the best of the situation and maximize his profit, or he could decline to cooperate and take something less. If he refused to cooperate, there was no hope of making up his financial loss; his only compensation was that the wholesaler was also punished for his greed.

When the game was played repeatedly, there was a significant difference. The retailer still had just two moves to use against a greedy wholesaler, but now his refusal to co-operate served the additional function of "educating" the wholesaler to lower his price in the future. And although he took an immediate loss as before, it was balanced by the prospect of larger future payoffs. Because of the retailer's greater bargaining power when the game was played repeatedly, the experimenter anticipated that there would be a stronger tendency to leave the Bowley point than when the game was played just once.

The second most significant variable was the amount of information available to the players. It was possible to have *complete information* or *partial information*. A player had complete information if he knew both his own and his partner's profits for all possible transactions. He had partial information if, for every possible transaction, he knew what his own profit would be but did not know his partner's profit. The experimenters felt that, as more information was made available to the players (especially to the retailer), a defection from the Bowley point to a Pareto optimal point became more likely. If the retailer had complete information, it was conjectured, he would be in a better position to see the disadvantages of a non-Pareto outcome and, in particular, the disadvantages of the Bowley point. A retailer who had only incomplete information, on the other hand, wouldn't know the difference between a point that was Pareto optimal and one that was not.

The least significant variable was the point of equal profit, the selling price and quantity of goods that allowed retailer and wholesaler to make the same profit. It was felt that, because of the cultural bias in favor of splitting profits equally, the Bowley point would be a more likely outcome if it was also an equal-profit point and a less likely outcome if the equal-profit point was Pareto optimal. (With the original prices mentioned earlier, the equal-profit point was reached when twenty items were sold for $10.10. The equal-profit point was Pareto optimal, and the Bowley point was not.)

To summarize, there were two basic choices: the Bowley point, at which each player tries to further his own immediate self-interest, and a Pareto-optimal outcome, at which the joint profits are maximized. There were three variables: the frequency with which the game is played, the amount of information available to the players, and the position of the equal-profit point. The purpose of the series of experiments was to determine the effect of various combinations of variables on the outcome.

Each of the three variables could involve either of two situations: there was complete or incomplete information; the game was played once or repeatedly; and the equal-profits point was Pareto optimal or at the Bowley point. There were eight possible combinations of variables in all. Actually, only five were used, because some of the combinations were of little or no value. If only partial information was available, for example, the equal-profits point would not be significant because it would not be known to the players. Of the five variations, there was only one case in which the players moved away from the Bowley point toward one that was Pareto optimal. That was when all the variables were favorable: there was complete information, the play was repeated, and the equal-profits point was Pareto optimal.

When the games were played repeatedly, a situation came up that was similar to one we talked about earlier in discussing the "prisoner's dilemma." Often wholesalers who initially set a high price were induced to come down when the retailer refused to buy. The two would then continue at this mutually agreeable price until the last trial, when the wholesaler would suddenly raise his price, putting the retailer in the dilemma he had faced in the one-trial game:

should he punish the wholesaler (and himself) for breaking faith, or take his small profit? What was done in each instance depended on the personality of the individual retailer. Interestingly enough, 50 percent of the subjects who were businessmen tried to squeeze a little extra out of the last trial by raising their price. When the subjects were students, however, deviations in the last trial were almost nonexistent.

The Personalities of the Players

There were many variations in the experimental results which could not be attributed to the three variables in the games. Apparently the outcome of the experiment was influenced considerably by what the players brought with them: their own personalities.

Fouraker and Siegel categorized the subjects as simple maximizer, rivalist, or cooperator. The simple maximizer was interested primarily in his own payoff. The rivalist was interested primarily in doing better than his partner; his own payoff was only secondary. And the cooperator was interested in helping both himself and his partner. This is not to say that invariably an individual adopted one of these points of view and stayed with it throughout the experiment. The experimenters felt that a predisposition to one of these points of view could be modified by appropriate payoffs. As a matter of fact, they showed that it could. When the players were offered a bonus if the joint profits exceeded a certain amount, cooperative behavior was induced. When a bonus was offered to the player who did better, rivalistic behavior was more prevalent.

The category to which each player belonged could be observed in several ways. Cooperative wholesalers, for example, would frequently quote prices that equalized the profits or that gave the retailer a large profit at the cost of a small loss to himself. Cooperative retailers ordered more items than they might have, if they had been interested only in their own profits, in order to raise the wholesaler's profit. When the wholesaler tried to raise his own profits considerably at a small cost to the retailer, some retailers punished him by ordering less. Others cooperated. In any case, it was clear that the experimental variables were insufficient to account for the differences in the players' behavior.

The biggest defect of experimental games is the inadequacy of the payoffs. In this respect, the series of experiments run by Fouraker and Siegel were better than most. They tried to restrict participation to people who would find the payoffs significant. Every player had answered an ad that offered a salary of a dollar an hour, and the average return to the players in the experiment was generally higher than this. Nevertheless, the *individual* payoffs for a single game were quite small, and undoubtedly this affected the play. If a wholesaler demanded 75 cents when the potential joint profit was a dollar, many subjects would be tempted to give him nothing even if the game was to be played only once and even if the subject was in sufficient need to be willing to work for a small wage. If the dollar were a million dollars and the player were offered a quarter of a million in a one-trial game, he would be much slower to take this cavalier attitude. In real "games," the stakes are frequently high enough so the problem doesn't arise.

A Bidding Experiment

Several years ago, James Griesmer and Martin Shubik ran some experiments using as their subjects Princeton undergraduates. The basic experiment went like this:

Two players simultaneously picked a number between 1 and 10. The player who picked the higher number received nothing. The player who picked the lower number received that amount (in dollars or dimes, depending on the experiment) from the *experimenter*. If both players picked the same number, they tossed a coin and the winner of the toss took that number. If, for example, one player picked 5 and the other picked 7, the player who picked 5 received 5 units from the experimenter, and the other player received nothing. If both players picked 5, a coin was tossed and the winner of the toss received 5 while his partner received nothing. After each trial, the players were informed what their partners had done, and the game was repeated. The players were separated throughout the series and could not consult.

This game is almost identical to the gasoline price war example discussed earlier. By the same induction argument used before, we can "deduce" the superiority of the most uncooperative strategy: a bid of 1. If both players adopt this strategy, the payoffs will be very small—on average 1/2. If one player is competitive and the other is not, the competitive player will generally win something, but not very much. The greatest rewards clearly go to the players who cooperate. If both players keep their bids high, the joint profit will be greatest, and though one of the players will necessarily get nothing on any particular trial, repetition of the game gives both players a chance to do well.

If both players in a pair realize that they will do best if they cooperate, they still have the problem of figuring out how to synchronize their bidding, since no direct communication is possible. The most direct way to cooperate, and the way that maximizes the joint expected profit, is to bid 10 all the time. This yields an average return of 5 to each player. The one flaw in this plan is that a player can still get nothing, since in the case of ties the payoff is determined by chance. At a small cost in expected value, the players can raise their assured profit considerably. On the first of each pair of trials, each player picks a 10. The player who won the toss on the first trial picks another 10 on the second trial, while his partner chooses a 9. Thus, on every pair of trials, one player always wins 10 and the other always wins 9. The average return to each player is 4 3/4, and the guaranteed return is 4 1/2.

As it happens, almost all the players approached the game competitively; they tried to outsmart one another rather than jointly exploit the experimenter. Some players made cooperative overtures by bidding 9's and 10's alternately, but these were most often misinterpreted as attempts to lull the partner into a false sense of security. (This was later confirmed by the subjects in explicit statements and was obvious from their behavior during the play.) A few of the pairs did manage to get together, and when they did, it was invariably by alternating bids of 9 and 10, so that each won 9 on every other trial. This is not the most efficient way to cooperate, of course, but it is fairly good for subjects playing for the first time.

One of the things the experimenters wanted to study was the "end effect"—uncooperative play on the last of a series of trials—but no evidence of it was observed during the experiment. When players managed to get together, they played cooperatively throughout the series of trials. If anything, there was more cooperation at the end than at the start. The experimenters tried to isolate end effect by telling some pairs the exact number of trials they would play and not telling others. Presumably, the end effect would occur when the players knew which was the last trial, but in fact that there was no difference in play. Perhaps the payoffs were too small to motivate defection. Generally, two players who "found each other" were so pleased that they left well enough alone. (When the players were uncooperative from the first—and most were—the situation didn't arise.)

One variation of the game is worth describing in detail because it illustrates a point we made before in connection with utility theory. In one experiment, the basic game was played in sets of three trials, with one additional rule: if in the first two games a player received nothing, in the third game he automatically received whatever he bid. This meant that a player could *always* get a payoff of 10 by bidding 10 on all three trials.

Players who were competitive in the earlier games usually played competitively in this game as well. Each of the players made low bids during the first two trials, and each won one of the first two games. The situation then was exactly the same as if they had been playing an ordinary game, since the new rule was not applicable. The players realized that if they played competitively on the last trial, as they had before, their profit, even if they won, would be considerably less than it would have been if they had played differently from the start. The realization that they could have done better apparently affected their utility functions, for they bid much much more in the third game than they had earlier—in what essentially were identical situations.

An Almost Cooperative Game

The last experiment we will discuss is a cooperative, two-person game in which the players have identical, or very similar, interests. The basic problem in this type of game is coordinating the players' strategies to their mutual advantage. In games in which the players cannot communicate directly, Thomas Schelling suggested, the players must look for certain clues to help them anticipate what their partners will do. The clue could be a past outcome, for example, or it might be suggested by the symmetry of the payoff matrix. Richard Willis and Myron Joseph ran experiments to test Schelling's theory that prominence was a major determinant of bargaining behavior. In all, three different matrices were used. We call them *A*, *B*, and *C* (see Figure 36).

The players were divided into two groups. Group *I* played game *A* repeatedly and then switched to game *B*, which they also played repeatedly. Group *II* started with game

B and ended with game C. No communication between the players was allowed during the play.

A

(10, 20)	(0, 0)
(0, 0)	(20, 10)

B

(10, 30)	(0, 0)	(0, 0)
(0, 0)	(20, 20)	(0, 0)
(0, 0)	(0, 0)	(30, 10)

C

(10, 40)	(0, 0)	(0, 0)	(0, 0)
(0, 0)	(20, 30)	(0, 0)	(0, 0)
(0, 0)	(0, 0)	(30, 20)	(0, 0)
(0, 0)	(0, 0)	(0, 0)	(40, 10)

Figure 36

Clearly, the players must solve the problem together. Unless they pick the same row and column, neither player will get anything. As a secondary goal, each player may try to reach the most favorable outcome of those that lie along the diagonal.

In game A, Schelling offers us no clue as to what the proper play should be. In game C, also, no clear-cut outcome suggests itself, but one of the two middle strategies would seem more likely than the extreme ones. It is only in game B that symmetry dictates a clear choice: the (20, 20) payoff corresponding to each player's second strategy.

What actually happened was quite surprising. When group I played game A, there was a battle of wills in which each player fought for dominance. Each player played the strategy that would give him a payoff of 20 if the other yielded. After the initial battle, in which there were few agreements, one player finally gave way and that team settled down to an equilibrium point. When they switched to game *B*, their play was strongly influenced by what had happened before. Rather than go to the (20, 20) payoff which symmetry would suggest, these players went most often to one of the asymmetric equilibrium points. *Three-quarters of the time, the player who was dominant in game* A *remained dominant in game* B *too.*

The actions of group *II* were even more unexpected. They started with game *B* and arrived at an agreeable equilibrium point much faster than group *I* had, which is not too surprising. But the agreement tended to be at one of the extremes —the first row and column or the third row and column— rather than on the middle row and column as suggested by symmetry and Professor Schelling. As before, when the players in group *II* switched to game *C*, the dominance that had been established in the earlier game prevailed.

Generally, repetitive play—picking the same strategy again and again—was the signal that was used to suggest an outcome. The repetitive play continued when it struck a responsive chord in the partner, and even, often, when it didn't. The most equitable outcome in games *A* and *C* would be a synchronized, alternating scheme in which one partner is favored in one play and the other is favored in the next. This never happened, however, probably because such an arrangement is too intricate to arrange without direct communication.

Some General Observations

The purpose of studying experimental games is to isolate the factors that determine how players behave. Ultimately, one hopes, enough knowledge will be obtained to allow us to predict behavior. So far, however, there has been only a limited amount of experimentation and the results have not been entirely consistent. Much of the work has centered on the two-person, non-zero-sum game, focusing on the "prisoner's dilemma," and in particular on identifying the elements that determine whether a person will compete or cooperate.

Participants in a "prisoner's dilemma" game, it has been established, often fail to cooperate. We mentioned several possible reasons for this earlier. A player may be interested only in furthering his own interests. He may be trying to do better than his partner (rather than maximizing his own payoff). He may feel that his partner will misunderstand his cooperative overtures or, even if he understands, may only exploit them. Or he may simply not understand the implications of what he is doing.

There is little doubt that players often fail to see all there is to see in a game. Professor Rapoport has suggested that people fail to play minimax in zero-sum games (and they frequently do) because of lack of insight. There is some evidence that people fail to cooperate in non-zero-sum games for the same reason. Often, players just do not conceive of playing other than competitively. In one set of "prisoner's dilemma" experiments, it was established by post-experimental interviews that, of twenty-nine subjects who understood the basic structure of the game, only two chose to play noncooperatively. In the bidding experiments conducted by Griesmer and Shubik which we discussed earlier, many of the players were competitive because they weren't aware of all the possibilities. They perceived the game as being competitive and never thought about cooperating. And those who did cooperate felt uneasy—as if they were cheating the experimenter by conspiring illegally.

Despite all this, and even granting the truth of Rapoport's observations on zero-sum games, something more than the ignorance of the players seems to be involved when players fail to cooperate in non-zero-sum games. For one thing, highly sophisticated people often play uncooperatively, especially in one-trial games. For another, if the rewards are sufficiently enticing, either cooperative or competitive behavior can be induced. Apparently the factors that determine whether the play will be cooperative or competitive are quite complex and the experiments do not indicate the ratio of distrust, ambition, ignorance, and competitiveness that leads to competitive play.

Not all the significant variables in a game are controlled by the experimenter. One of the most important elements of the game is the personality of the player. It is to be expected that two people will react differently even in identical situations. If we are to have any chance of predicting the outcome of a game, therefore, we must look beyond the formal rules and into the attitudes of the players—a difficult job.

Experimenters have long been aware of the importance of personality and have tried to measure it or control its effect in a number of ways. The following clever scheme, for example, was devised by Jeremy Stone to measure the aspirations (or greediness) of a player and his attitude toward risk.

A player was given a large number of cards and was told that he would be playing a series of games against an unnamed opponent. On each card were listed the rules of a particular game. A typical card might read: "You and your partner will each pick a number. If your number plus twice your partner's does not exceed 20, you will each get the number of dollars that you picked; otherwise, you will each get nothing." Among the cards there would also be a game exactly like this one, but with the players' roles reversed. It would read: "You and your partner will each pick a number. If twice your number plus your partner's does not exceed 20, you will each get the number of dollars that you picked; otherwise, you each get nothing." The cards were then paired, so that the subject played the role of both players—in effect, playing against himself. The final score of a player was dependent on his own attitudes only.

Some experimenters tried, but failed, to correlate what a player did in the "prisoner's dilemma" with his scores on certain psychological tests. Some connection *was* found between the players' political attitudes and how they played. In one study by Daniel R. Lutzker, players were scored on an internationalism scale on the basis of how they responded to statements such as: "We should have a world government with the power to make laws which would be binding on all member nations," or "The United States should not trade with any Communist country." Lutzker observed a tendency of extreme isolationists to cooperate less in the "prisoner's dilemma" than did extreme internationalists.

Morton Deutsch, in a similar study, found that two characteristics related to the "prisoner's dilemma" were dependent on each other, and each of these, in turn, was dependent on the person's score on a particular psychological test. The traits Deutsch was concerned with are "trustingness" and "trustworthiness." To study these, a variation of the "prisoner's dilemma" was tried in which the play was broken down into two stages. In the first stage, one player picked his strategy. In the second stage, the other player picked his strategy, *after being informed what the first player had done*. The subjects played both roles, and in many different games. A player was *trusting* to the extent that he played cooperatively when he played first; he was *trustworthy* to the extent that he played cooperatively at the second turn in response to a cooperative play by his partner. It turned out that, to a significant degree, "trustingness" and "trustworthiness" are positively correlated, and each trait is inversely related to authoritarianism (as measured by a certain psychological test: the F scale of the Minnesota Multiphasic Personality Inventory [MMPI]).

There have been attempts to relate other attributes of a person, such as his intelligence or sex, to the way he plays, but with little or no success. The indications are that a person's previous experience—and, in particular, his profession—affects his play. We have already mentioned how differently businessmen and students behaved in the last trial of a series of wholesaler-retailer games. Incidentally, a study was made of the "prisoner's dilemma" using prisoners as subjects. The results seem to contradict the adage about honor among thieves. Indeed, prisoners behaved very much like college students when playing this game: by and large, they tended not to cooperate.

Experimenters have also tried to control personality in games by imposing an attitude on the players. The players were told: "Do the best for yourself and don't worry about your partner," or: "Beat your partner." In this way, it was hoped, the players' attitudes would be fixed. These instructions had little effect, however.

Behavior Patterns

Certain consistent patterns have been observed by several experimenters who have studied sequences of "prisoner's dilemma" games. There is a tendency for players to become less cooperative rather than more cooperative as the games are repeated—though it is not clear why. Also, the reaction of subjects to their partner's behavior is in a way consistent. Suppose a sequence of "prisoner's dilemmas" is divided into two halves. Suppose, also, that in one case a player's partner always acts cooperatively in the first half, and in another case the partner always acts noncooperatively in the first half. (In such experiments, the "partner" is either the experimenter or an assistant.) Noncooperative playing in the first half of the games, it was found, is more likely to elicit a cooperative response in the second half than is cooperative playing.

The effect of allowing the players to communicate is also subtler than might at first appear. Some experimenters found that the chance of a cooperative outcome was increased if the players were allowed to communicate. An experiment by Deutsch indicated, however, that this was so only for individualistic players: players who wanted to win as much as possible and who didn't care how their partners did. The cooperative player who was also concerned about how his partner did, and the competitive player whose main concern was doing better than his partner, played the same way, whether they were able to communicate or not.

In another experiment carried out by Deutsch to analyze further the effects of communication, three basic situations were set up: a bilateral threat, in which each player had a threat against the other; a unilateral threat, in which only one player had a threat; and a no threat. (A person has a threat if he can lower his partner's payoff without changing his own.) In each of these three situations, some players were given the opportunity to communicate and others were not. Players who had the opportunity to communicate chose not to do so, however. In one situation—the bilateral threat—there was more cooperation when the players could communicate but didn't than when they did. When the players did try to communicate, negotiations often degenerated into repetition of threats. In the unilateral case, on the other hand, if the players communicated they generally became less competitive; and when the players could communicate but didn't, conflicts between them were created or aggravated. In short, the effect of allowing communication depends on the attitudes of the players, and, in turn, the attitudes of the players may be affected by the ability to communicate.

The most significant drawback to these experiments is, again, the insignificance of the payoffs. This was manifested explicitly and implicitly in many ways. It was reflected in the increased competition: beating one's partner became more important than maximizing one's payoff. It was indicated by the players' own statements; they admitted playing frivolously during long, monotonous, "prisoner's dilemma" runs. It was shown by the general tendency (at least on the part of students) to be cooperative in the wholesaler-retailer game once a tacit understanding had been reached and by the speed with which defectors were punished. (It is easier to punish a greedy partner when it costs you $10 than when it costs $10,000.) Our point is not that the experimenters erred, but rather that the results must be analyzed cautiously.

The Two-Person, Non-Zero-Sum Game in Practice

It is difficult to study "games" that are actually "played" in real life. Still, it has been tried. Determining how fruitful these attempts have been is better left to the experts. We will be content to say just a few words in passing.

One of the fields in which game theory has been studied is business, and the analysis has been essentially descriptive. One study which we referred to earlier concerned the tactics used in a taxicab rate war. Elsewhere, competition in business was compared with armed conflict, and the conditions that tend to precipitate price wars were pinpointed. In the field of advertising, more formal models have been constructed, encompassing many problems: setting the advertising budget, allocating funds to the various media and/or geographic areas, determining the best time slots to advertise in, etc. There has even been an attempt to fit Indian burlap trade into the framework of game theory, allowing for possible coalitions, strategies, and threats.

Political scientists have also borrowed concepts from two-person, non-zero-sum game theory. We mentioned Professor Maschler's application to disarmament models. Other models have centered about nuclear deterrence and bomb testing. Some models have been studied in extensive form, with the "plays" being military resources: transforming non-military resources into military resources, standing pat, etc. The role played by communication, especially between the United States and Soviet Russia, has also been studied on this basis.

In these models, it is virtually impossible to get an accurate quantitative expression which will reflect the payoffs. Still, enough insight may be obtained by assigning approximate figures to justify the studies.

6

The n-Person Game

The New York Times Magazine of October 20, 1968, carried an article entitled "The Ox-Cart Way We Pick a Space-Age President," in which the role played by the electoral college in Presidential elections was examined. The author asserted that voters in large states have an advantage over voters in smaller states, despite the uniform bonus of two senatorial votes which is allotted to large and small states alike. In the words of John A. Hamilton: "Under the winner-take-all methods of allocating electoral votes the big states exercise inordinate power. Although many more voters will go to the polls in New York than in Alaska, they will have a chance to influence many more electoral votes and, on balance, each New Yorker will have a far better chance to influence the election's outcome."

The advantage that the voter from a large state has is not

immediately obvious. The bloc of electoral votes controlled by New York is clearly more powerful than that of Rhode Island, but an individual voter in New York has less influence on how his state will vote than the individual voter in Rhode Island does, and it is the strength of the individual voter with which we are concerned. Hamilton balances these two opposing tendencies and comes up with a plus for the large states. For evidence, he looks to the record. He points out that a switch of only 575 popular votes in New York in 1884 would have made James Blaine President instead of Grover Cleveland. He then lists four Presidential elections in which 2,555 votes, 7,189 votes, and 10,517 votes in New York and 1,983 votes in California would have made the difference. On the basis of this he concludes that the large states have an influence on Presidential politics which surpasses their actual size.

Granted that large states (or, rather, voters in large states) have the advantage in the electoral college, the question arises whether this is a peculiarity of the way the votes are distributed in our Congress or is always true in this kind of voting system. In general, how can one assess the "power" of an individual voter?

The fact is that voters in large states are not necessarily more powerful than other voters. Suppose there are five states, four with a million voters and one with 10,000 voters. Now, whatever representation the state with 10,000 voters has in the electoral college (as long as it has *some* representation), it is as strong as any other state, since a majority must include three states—*any three states.* In effect, all states have equal power in the electoral college (assuming decisions are by majority vote and the large states all have the same representation). But an individual voter in a small state has much more influence on how his state will vote, and consequently is more powerful.

Suppose we modify the situation slightly. Assume that one state has 120 votes in the electoral college, three states have 100 votes, and one state has 10 votes. Now the 10-vote state is absolutely powerless, and so are the state's voters. No matter what the final vote is, the position of the 10-vote state is irrelevant, since it cannot affect the outcome by changing its vote; the voters in the small state are completely disfranchised.

The second problem—that is, actually assigning numbers to the players which will reflect their voting strengths—we will consider later.

The Presidential election is itself a kind of game. The voters are the players, the candidates are the strategies, and the payoff is the election results. This is an n-person game, a game in which there are more than two players. The distinction is convenient, for games with three or more players generally differ radically in character from those with fewer than three players.

Of course, there may be considerable variation in n-person games also. Some examples follow.

Some Political Examples

In a local political election, the main issue is the size of the annual budget. It is well known that each voter prefers a certain budget and will vote for the party that takes a position closest to his own. The three parties, for their part, are completely opportunistic. They know the voters' wishes and want to gain as many votes as possible; they couldn't care less about the issues. If the parties have to announce their preferred budgets simultaneously, what should their strategy be? Would it make a difference if their decisions were not made simultaneously? Would it make much difference if there were ten parties instead of three? Two parties?

In 1939, three counties had to decide how some financial aid, allocated by the state, would be divided among them. This aid was to go toward school construction. In all, four different schools were to be built and distributed among the three counties.

Four different plans were proposed—we will call them *A*, *B*, *C*, and *D*—each prescribing a different distribution. The final plan was to be selected by a majority vote of two of the three counties. The relationship of the counties, the plans, and the distribution of the schools are shown in Figure 37.

		Plan			
		A	*B*	*C*	*D*
	I	4	1	2	0
	II	0	0	1	2
County	*III*	0	3	1	2

Figure 37

The numbers in the table indicate how many schools will be built in each county according to each plan. Assuming that the plans are voted on two at a time until all but one is eliminated, which plan will win if each county backs the plan that gives it the most schools? Does it matter in what order the vote is taken? Would it pay for a county to vote for one plan when it actually prefers another?

A nation that elects its governing body by a system of proportional representation has five political parties. Their strengths in the legislature are determined by the number of seats they hold: eight, seven, four, three, and one. A governing coalition is formed when enough parties join to ensure a majority; that is, when the coalition controls twelve of the twenty-three seats. Assuming that there are benefits that will accrue to the governing coalition—such as patronage, ministerial portfolios, etc.—what coalition should form? What share of the spoils should each party get?

Some Economic Examples

In a certain city, all the houses lie along a single road. If we assume that customers always buy from the store nearest them, where should the retailers build their stores? Does the number of stores make any difference?

Several fashion designers announce their new styles on a given date, and, once announced, they can't be changed. The most important feature of the new dresses and the feature that determines which dress is finally bought is the height of the hemline. Each customer has her own preference and will buy the dress that best approximates it. Assuming the designers know the percentage of women who will buy each style, which styles should the companies manufacture?

An agent writes three actors that he has a job for two of them, any two of them. The three actors are not equally famous, so the employer is willing to pay more for some combinations than he is for others. Specifically, A and B can get $6,000; A and C can get $8,000; and B and C can get $10,000. The two that get the job can divide the sum any way they like, but before they can take the job, they must decide how to divide the money. The first two actors to reach an agreement get the job. Is it possible to predict which pair will get the job? How will they divide the profits?

A wholesaler wants to merge with any one of four retailers who jointly occupy a city block. If the merger goes through, the wholesaler and the retailer will make a combined profit of a million dollars. The retailers have an alternative: they can band together and sell to a realty company, making a joint profit of a million dollars that way. Can the outcome be predicted? If the wholesaler joins a retailer, how should they divide the million dollars?

An inventor and either of two competing manufacturers can make a million dollars using the patent of one and the facilities of the other. If the inventor and one of the manufacturers should manage to get together, how should they share their profit?

An Analysis

One of the advantages of using a game as a model is that it permits you to analyze many apparently different problems at the same time. That is the case here. In the local political election, for instance, the parties are in a very similar situation as the retailers in the first economic example, and they in turn have the same problem as the fashion designers in the second economic example. In terms of the political example, the issue need not be quantitative, such as fixing a budget. It might involve taking a position on any problem where there is a more or less one-dimensional spectrum of opinion, such as liberal vs. conservative, or high tariff vs. free trade. At first, it might seem that the parties should take a position somewhere in the middle; and if there are only two parties, and they make their decisions in sequence rather than simultaneously, this is generally what happens. In Presidential elections, the conservative party and the liberal party often (though not invariably) tend to move to the center, on the theory that voters on the extremes have nowhere to go. Under similar circumstances, the competing stores will tend to cluster at the center of town; and the fashion designers will follow moderate tastes. When there are many political parties (or many stores in town, or many fashion designers), it may be better to cater to some outlying position. If there is proportional representation, one of ten parties may be willing to settle for 20 percent of the vote, even if it means giving up any chance of getting the votes at the center.

If you look at the problems that the players must solve in these games—how to win the most votes, how to get the most schools, how to sell the most dresses—you see that the primary concern is with power: the power of a player or a coalition of players to affect the final outcome of the game.

Power in the context of the n-person game, while real

enough, is a subtler concept and more difficult to assess than in the simpler, one-person and two-person games. In the one-person game, a player determines the outcome himself or shares control with a non-malevolent nature. In the two-person, zero-sum game, a player's power—what he is sure of getting on the basis of his own resources only—is a good measure of what he can expect from the game.

The two-person, non-zero-sum game is a little more complicated. Here, a player also wields the power to punish or reward his partner. Whether this power over the partner can be made to increase the player's own payoff, and to what extent, depends on the partner's personality. Since it cannot be so converted by the player himself, its value is not completely clear. Nevertheless, this kind of power is highly significant and must be taken into account in any meaningful theory.

In the n-person game, the concept of power is even more elusive. Of course, there is always a minimum payoff which the player can get by himself. To get any more, he must join others, as he did in the two-person, non-zero-sum game. But in the n-person game, if other players fail to cooperate, he has no recourse. It would seem, then, that beyond the minimum payoff he is helpless. Yet coalitions of apparently impotent players do have power. Thus, the player has potential power which requires the cooperation of others to be realized. Making this concept of potential power precise is one of the basic functions of n-person game theory.

To illustrate how the problem of evaluating a player's potential power might be approached, let us look at one of our earlier examples. Suppose that, in the third economic example, one of the actors approaches an outside party before the bargaining begins and offers him whatever earnings he receives in return for a lump sum. The third party would take over the bargaining for the actor: he could offer and accept any bargain he wished on the actor's behalf and, of course, would take the risk of being left out completely, with no gain at all. The amount that the third party would be willing to pay for the privilege of representing the actor might be considered an index to the actor's power in the game.

This states in a more specific way the problem of evaluating potential power but takes us no closer to a solution. Actually, many different solution concepts are possible for this kind of game, and we shall discuss them later. For the moment, let us look at one approach. The sketch in Figure 38 summarizes the game.

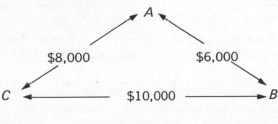

Figure 38

In this game, one's first inclination is to assume that B and C will join together, since they have the most to gain: $10,000. Just how they will divide this money is another matter. One suspects that player A, who is not even included in the coalition of B and C, will play an important part in determining how the money is divided, for if B and C should fail to reach an agreement, they would both have to look to A. The respective shares of B and C should be somehow related to the value of the coalitions AB and AC. And since the coalition AB has a smaller value than the coalition AC, it seems reasonable to conclude that B will get less than half of the $10,000.

This argument has an obvious weakness, however. Once A realizes that he has an inferior position, he is bound to lower his demands. It is clearly better for him to get into some coalition, even if he has to leave the lion's share to his partner, than to stand by himself and get nothing. Thus, even the one "obvious" conclusion—that B and C will form the coalition—now seems open to question.

This should give a rough, qualitative idea of how one might attack the problem, and that is enough for now. It might be interesting to anticipate a little, however, and look at some of the conclusions reached about this game on the basis of one game theory, without dwelling on the reasoning behind the conclusions. We will use the Aumann–Maschler theory, because it is particularly simple.

In the three-actor example, the Aumann–Maschler theory does not predict which, if any, coalition will form. It does predict, however, what a player will get if he manages to enter a coalition; and this amount, at least for this game, does not depend upon which coalition forms. Specifically, the theory predicts that A should get $2,000; B, $4,000; and C, $6,000.

In the wholesaler-retailer merger, the Aumann–Maschler theory does not predict what coalition will form (it never does). In fact, in this case it does not even predict the precise way the players will divide the million dollars. It simply states that if a retailer joins the wholesaler, then the retailer will get something between a quarter of a million and a half million dollars. In the problem of the inventor and the two competing manufacturers, the theory predicts that the inventor will get virtually all the profit.

The school-construction and the national-election examples are two more illustrations of voting games—a common source of n-person games. The problem of assigning strengths to the parties in the legislature is very similar to that of assigning strengths to states in the electoral college (a problem we discussed earlier). In the national-election example, we have a situation in which the real power of the players (that is, the parties) is not what it seems to be. In this instance, the party with four seats has exactly the same strength as the party with seven seats. The party with four seats, if it is to be a member of a majority coalition, must combine either with the eight-seat party (and possibly others), or with the seven-seat party and a smaller party (and possibly others). So must the seven-seat party (except that, in the second alternative the roles of the seven-seat and four-seat parties are reversed).

Finally, in the school-construction issue, the order in which the counties vote is critical. Suppose the order is *CABD*. This means that proposal *A* and *C* are voted on first; then *B* is pitted against the winner of that contest; and finally *D* is matched against the winner in the second contest. The outcome may be depicted as shown in Figure 39. Thus, in an election between *C* and *B*, *C* would win;

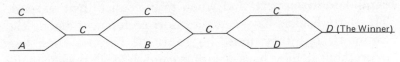

Figure 39

this may be indicated by $C > B$. Also, $D > A$, $C > A$, $D > C$, $B > D$, and $A = B$. (That is, there would be a tie between *A* and *B*.)

If the order is *DACB*, the outcome is as shown in Figure 40. Note that in the first order of voting, *CABD*, it is to

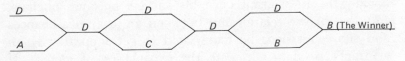

Figure 40

county *I*'s advantage to vote strategically. In the second contest, county *I* should vote for *B* rather than for *C*, even though it prefers *C*. If it does vote for *B*, *B* will win the second vote and in the last contest *B* will win again. In this way, county *I* gets one school. If county *I* always voted blindly for the plan it preferred without regard for strategic considerations, plan *D* would be adopted and county *I* would have no schools at all.

Admittedly, these illustrations are highly simplified. Reality is, almost invariably, much more complicated. One senator, for example, may be much more powerful than another, despite the fact that each has one vote. A phenomenon as complex as the election of the President of the United States is almost impossible to capture in a single model. Consider some of the difficulties.

In the process of electing a President, many games are played. One of them takes place at the party national convention. Among the delegates are a small number of party leaders who control a large percentage of the votes and who act (they hope) so as to maximize their political power. These party leaders must make two basic decisions: which candidate to support, and when to announce their support. The party leaders have some a priori preferences among the aspirants, but they must be cautious, because they lose everything if they back a losing candidate. If they manage to back the eventual winner, they reap certain benefits, depending on the strength of the leader, the need of the candidate for their support, etc. These benefits may take the form of reciprocal support or patronage.

Two political scientists, Nelson W. Polsby and Aaron B. Wildavsky, constructed a model of a convention that took into account such phenomena as (1) the band wagon flocking to an aspirant once his success becomes obvious; (2) the inverse relationship between an aspirant's chances and the concessions he must yield to get support; (3) an aspirant's need to stimulate the expectation of victory. If an aspirant doesn't win or increase his vote within a certain period of time, his vote thereafter will not remain constant; it will drop. As a result, aspirants often "hide" votes they control and release them slowly when they think they'll do the most good.

Once the candidates are nominated, there is a new game —very similar to the one described in our first political example. Now the two parties must take a position not on a single issue but on many. Of course, they are not quite as free in doing so as we assumed in our example, since parties do have ideological commitments.

The voter, too, plays a game. Even in the simplest

instance, where there are only two candidates, there may be problems. If a voter's position is at one end of the political spectrum, his wishes may be virtually ignored if he votes mechanically for the candidate whose position is closest to his own. If he withholds his vote, the election of the other, less favored, candidate becomes more likely. Balanced against this risk is the chance of increasing his influence in the future. The political parties will take the extreme voters more seriously if, whenever they move too far toward the center, there are wholesale defections. (When there are more than two serious candidates, the situation is even more complicated, for then you must decide whether to vote for a very desirable candidate with a doubtful chance of winning or for a less desirable candidate with a somewhat better chance of winning.)

Obviously, it is out of the question to analyze all these games in their original form. Instead, we must take simple models and try to formulate a solution that seems plausible. This may be done, and in fact has been done, in several ways. Some solutions aim at reaching an arbitration point based on the strengths of the players. Others try to find equilibrium points such as exist in a market of buyers and sellers. Some define a solution as sets of possible outcomes satisfying certain stability requirements.

The von Neumann–Morgenstern Theory

In their book, *The Theory of Games and Economic Behavior,* von Neumann and Morgenstern first defined the n-person game and introduced their concept of a solution. All the work that has been done on n-person games since has been strongly influenced by this now classic work. It will be easier to understand the von Neumann–Morgenstern approach (from now on we will refer to it by the abbreviated N-M) if we discuss it in the context of a specific example.

Suppose that three companies—*A*, *B*, and *C*—are each worth a dollar. Suppose that any two, or all three of them, can form a coalition. If such a coalition forms, it will obtain an additional $9, so that a two-person coalition will be worth $11: the dollar each company had originally, plus the additional $9. And the three-person coalition will be worth $12. We assume that each company is completely informed and, for the sake of simplicity, that utility is identical to money. What remains is to determine the coalition that will form and the way the money will be divided. Before attacking this problem, however, we will make some general observations about n-person games.

The Characteristic Function Form

The game which we just described is said to be in *characteristic function form*. With each coalition is associated a number: the value of that coalition. The value of a coalition is analogous to the value of a game in the two-person case. It is the minimum amount that the coalition can obtain if all its members join together and play as a team.

For many games, the characteristic function form is the most natural description. In a legislative body, for example, in which decisions are made by majority vote, the coalition values are obvious. A coalition that contains a majority of the players has all the power; a coalition without a majority has none. In other games—games in which the players are buyers and sellers in the open market, for instance—the value of a coalition may not be so clear. But N-M show that in principle this kind of game may also be reduced to characteristic function form, as follows.

N-M start with an n-person game in normal form. This is a game in which each player picks one of several alternatives, and in consequence of these choices, there is some outcome: a payoff to each of the players. The strategies available to the players may be fixing a price or quantity, casting a vote, hiring a number of new salesmen, etc. Having thus set the stage, N-M ask what would happen if a coalition of players—let us call them S—decide to act in unison to get the biggest possible combined payoff they can get. What should the coalition S hope to obtain?

This problem, N-M observe, is really the same problem we faced in the two-person game. The members of S constitute one "player," and everyone else constitutes the other "player." As before, we may compute coalition S's maximum payoff, assuming that the players not in S act in a hostile manner. This figure, denoted by $V(S)$, is called the *value* of coalition S; the value of any coalition may be calculated in this way.

This procedure suggests the same question that was raised before: Will the players not in S really try to minimize the payoffs of the players in S? And N-M's answer is the same here as it is in the two-person game: they will if the game is purely competitive. For this reason, N-M assume that the n-person game is zero-sum; that is, if the value of any coalition S is added to the value of the coalition consisting of the players not in S, the sum will always be the same. (If more than two coalitions form, the sum of the coalition values may decrease, but it will never increase.)

Superadditivity

Since there are many different kinds of n-person games, the values assigned to the coalitions may take on almost any pattern—almost, but not quite. A basic relationship exists between the values of certain coalitions which is a consequence of the way these values are defined.

Suppose R and S are two coalitions that have no players in common. A new coalition is formed which is composed of all the players either in R or in S; the new coalition is denoted by RuS (R union S). Clearly, the value of the new coalition must be at least as great as the sum of the values of coalition R and coalition S. The members of R can play the strategy that guarantees them $V(R)$, and the players in S can play that strategy that guarantees them $V(S)$. Thus, $R \cup S$ can get at least $V(R) + V(S)$. (It is quite possible, of course, that $R \cup S$ can do even better.) This requirement which must be satisfied by the characteristic function is called *superadditivity*. Stated another way, a characteristic function is superadditive if, for any two coalitions R and S which have no players in common, $V(R) + V(S) \leqq V(R \cup S)$.

Returning now to the original example, consider some of the outcomes. One possibility is that all three players will unite. In that case, symmetry would suggest that each of the players receive a payoff of 4. We will denote such a payoff by (4, 4, 4), the numbers in parentheses representing the payoffs to companies A, B, and C, respectively. Another possibility is that only two players will combine—say B and C—and share their $11 equally, giving the third player, A, nothing. In this case, the payoff would be (1, 5 1/2, 5 1/2), for A would still have $1. A third possibility is that the players are not able to come to an agreement and remain as they were originally, the payoff being (1, 1, 1). To see whether any of these outcomes, or some other outcome, is likely to materialize, let us imagine how the negotiations might go.

Suppose someone starts by proposing a payoff of (4, 4, 4). This seems fair enough. But some enterprising player, say A, realizes that he can do better by joining another player, say B, and sharing the extra profit with him. The payoff would then be (5 1/2, 5 1/2, 1). This is a plausible change. Both A and B would get more than they do in the earlier (4, 4, 4) payoff. C will be unhappy, of course, but there isn't much he can do about it—at least, not directly. But C can make a counteroffer. He might single out B and offer him $6, take $5 for himself, and leave A with $1—for a payoff of (1, 6, 5). If B accepts C's counteroffer, it is then A's turn to fight for his place in the sun.

Of course, this hopping about from payoff to payoff can be endless, for every payoff is unstable in that, no matter what payoff is being considered, there are always two players who have the power and motivation to move on to another, better, payoff. For every payoff there are always two players who together get no more than $8; these two players can combine and increase their joint profit to $11. Obviously, then, this approach doesn't work.

Imputations and Individual Rationality

When one is first introduced to n-person games, the temptation is to look for a best strategy (or a best set of equivalent strategies) for each player, and a unique set of payoffs which one might expect clever players to obtain; in short, a theory much like that of the two-person, zero-sum game. But it soon becomes clear that this is much too ambitious. Even the simplest games are too complex to permit a single payoff. And if you did construct a theory that predicted such a payoff, it would not be plausible or a true reflection of reality, since there are usually a variety of possible outcomes whenever an actual game is played. This is so no matter how sophisticated the players. There are simply too many variables—the bargaining abilities of the players, the norms of society, etc.—for the formal theory to accommodate.

One thing we can do, however, is to pare down the number of possible payoffs by eliminating those that clearly wouldn't materialize. This is what N-M do first. The N-M theory assumes that the ultimate payoff will be *Pareto optimal*. (An imputation, or payoff, is Pareto optimal if there is no other payoff in which all the players simultaneously do better.) On the face of it, this seems reasonable. Why should the players accept an imputation of (1, 1, 1) when all three players do better with an imputation of (4, 4, 4)? N-M also assume that the final imputation will be *individually rational*.

That is, each player will obtain, in the final imputation, at least as much as he could get by going it alone. In our example, this would mean that each player must get at least 1.

Domination

Let's go back to the original bargaining, but now we will assume that the only proposals to be considered are individually rational, Pareto optimal imputations. If a proposal is made—that is, a coalition along with an associated payoff —under what conditions will an alternative proposal be accepted in place of the original?

The first requirement is that there is a group of players strong enough to implement the alternative proposal—an agreement to bell the cat is worthless if there's no one around capable of doing it. In addition, the players who are to implement the new proposal must be properly motivated. This means that every player must get more than he would if he stuck with the old proposal. If both these conditions are satisfied, if there is a coalition of players with both the ability and the will to adopt the new proposal, we say that the new proposal *dominates* the old one and call the implementing coalition the *effective set*.

To see how this works in our original example, suppose the original coalition consists of all three players, with a payoff of $(5, 4, 3)$. The alternative payoff $(3, 5, 4)$ is preferred by B and C since they each get an extra dollar. Since B and C can get as much as \$11 acting together, they can enforce the new payoff. (They could, of course, limit A's payoff to 1, but they don't have to.) So $(3, 5, 4)$ dominates $(5, 4, 3)$, with B and C as the effective set. On the other hand, if $(1, 8, 3)$ were the alternative payoff to $(5, 4, 3)$, it would not be accepted. Any two of the three players have the power to enforce this payoff, but no two are so inclined. A and C both prefer the original payoff, and B, who wants to change, doesn't have the power to bring it off alone. An alternative payoff of (6, 6, 0) would be preferred to the original by both A and B. But in order to get a payoff of $12, all three players must agree, and C is not about to. C can get 1 by going it alone and is not likely to settle for less.

A convenient way to illustrate the imputations is based on an interesting geometric fact about equilateral triangles: for any two interior points the sum of the distances to the three sides is always the same. In our example, a payoff is an imputation if everyone obtains at least 1 and the sum of the payoffs is 12. We can represent all possible imputations, then, by points in an equilateral triangle which are at least one unit from each side. In the diagram in Figure 41, the

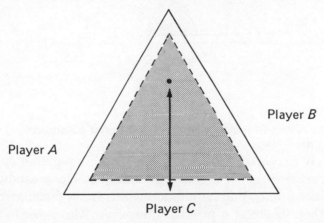

Figure 41

imputations are represented by the points in the shaded region, and the point P represents the imputation (2, 2, 8).

In the diagram shown in Figure 42, the point Q represents the payoff (3, 4, 5). In the horizontally shaded area are all the imputations that dominate Q with effective set BC. (The farther away an imputation is from the side marked "Player A," the greater the payoff to A in that imputation.) Both B and C get more in any imputation in the horizontally shaded area than they get at Q. The vertically shaded area and the diagonally shaded area represent those imputations which dominate Q with effective sets AC and AB, respectively. The plain areas contain the imputations that Q dominates, and the boundary lines represent the imputations that neither dominate nor are dominated by Q.

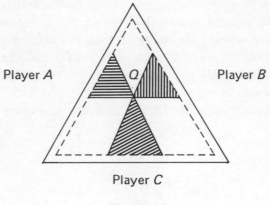

Figure 42

The von Neumann–Morgenstern Concept of a Solution

If you were asked to pick out a single imputation as the predicted outcome of a game, the most attractive candidate would seem to be that imputation which is not dominated by any other. There is a problem, however. There needn't be just one undominated imputation; there may be many. Worse still, there may be no undominated imputation at all. This is the case in our example. Every imputation there is dominated by many others. In fact, the domination relation is what mathematicians call intransitive. Imputation P may dominate imputation Q, which in turn may dominate imputation R. And imputation R may dominate imputation P. (Of course, the effective sets must be different each time.) That is why the negotiations went round and round, without settling anywhere.

From the start, N-M gave up any hope of finding a single-payoff solution for all n-person games. There might be particular games in which such a solution would be plausible, but "the structure . . . under consideration would then be extremely simple: there would exist an absolute state of equilibrium in which the quantitative shares of every participant would be precisely determined. It will be seen, however, that such a solution, possessing all necessary properties, does not exist in general."

After ruling out the possibility of finding a single, satisfactory outcome for all n-person games, N-M assert that the only reasonable outcomes are imputations and go on to define their concept of a solution. "This consists of not setting up one rigid system of apportionment, i.e., imputation, but rather a variety of alternatives, which will probably all express some general principles but nevertheless differ among themselves in many particular respects. This system of imputations describes the 'established order of society' or 'acceptable standard of behavior.' "

A solution, then, consists of not one but many imputations which together have a certain internal consistency. In particular, a solution S is some set of imputations which have two essential properties: (1) No imputation in the solution is dominated by any other imputation *in the solution*. (2) Every imputation that is not in the solution is dominated by an imputation that is in the solution.

This definition of a solution "expresses the fact that the standard of behavior is free from inner contradictions: no imputation y belonging to S [the solution]—i.e., conforming with the 'accepted standard of behavior'—can be upset —i.e., dominated—by another imputation x of the same kind." On the other hand, "the 'standard of behavior' can be used to discredit any non-conforming procedure: every imputation y not belonging to S can be upset—i.e., dominated—by an imputation x belonging to S." "Thus our solutions S correspond to such 'standards of behavior' as have an inner stability: once they are generally accepted they overrule everything else and no part of them can be overruled within the limits of accepted standards."

In general, there are many different solutions to any particular n-person game, and N-M do not try to single out a "best" one. The existence of many solutions, they feel, far from being a defect of the theory, is in fact an indication that the theory has the flexibility necessary to deal with the wide diversity one encounters in real life.

A related question—do solutions always exist?—is more serious. For N-M, the question was crucial. "There can be, of course, no concessions as regards existence. If it should turn out that our requirements concerning a solution S are, in any special case, unfulfillable—this would necessitate a fundamental change in the theory. Thus a general proof of the existence of solutions S for all particular cases is most desirable. It will appear from our subsequent investigations that this proof has not yet been carried out in full generality but that in all cases considered so far solutions were found." Since this was written, there have been many attempts to prove that solutions exist for all n-person games. They were all fruitless until 1967, when William F. Lucas structured a ten-person game for which there was no solution, and the twenty-year-old question was finally settled.

N-M's concept of a solution can most easily be explained by means of an example. Suppose A, B, and C are players in a three-person game in which any coalition with either two or three players can get 2 units, and a player alone gets nothing. This game has many solutions—an infinite number, in fact—but we will look at just two of them.

The first solution, which consists of only three imputations: $(1, 1, 0)$, $(1, 0, 1)$, and $(0, 1, 1)$, is indicated in the diagram in Figure 43. In order to prove that these three imputations taken together are really a solution, two things must be verified: there must be no domination between imputations in the "solution," and every imputation outside the "solution" must be dominated by an imputation within it. The first part is easy enough. In passing from one imputation in the "solution" to another, one player always gains

1, one player always loses 1, and one player stays the same. Because a single player is not effective, there can be no domination. (In order to be in the effective set, a player must gain in the change. It is not enough that he not lose.)

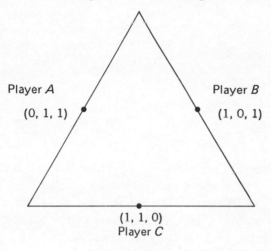

Figure 43

Also, in any imputation outside the "solution," there would be two players who get less than 1. This follows from the fact that no player gets a negative payoff, and the sum of the payoffs is 2. It follows that this payoff would be dominated by the payoff in the "solution" in which both these players received 1. The two players who originally received less than 1 would of course be the effective set. This shows that the "solution" is in fact a solution.

Another solution would consist of all imputations in which one player—say, player A—received $1/2$. We leave the verification of this to the reader. Figure 44 shows the solution.

The first solution may be interpreted in the following way. In every case, two players will get together, divide their 2 equally, and leave nothing for the third player. (Nothing is said about *which* two will join, however.) The payoffs are, of course, Pareto optimal. (They are also the most efficient, in that the gain per player is 1, while in a coalition of three the gain per player would be only 2/3. They are also enforceable. This is called the symmetric solution, because all the players have identical roles.

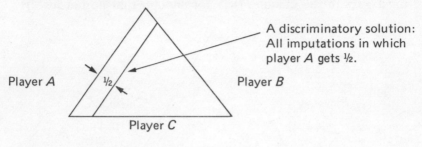

Figure 44

In the second solution—a discriminatory solution—two players join, give the third player something less than his "fair share" of 2/3, and take the rest for themselves. Which of the players join, and what is given to the third player, is determined by such factors as tradition, charity, fear of revolution, etc. Once the amount to be given to the "outsider" is determined, the game degenerates into an essentially two-person bargaining game, with the outcome depending on the personality of the players, and therefore indeterminate. All possible divisions between the two players, then, are included in the solution.

Some Final Comments on the N-M Theory

In order to construct a theoretical model of a real game, it is generally necessary to make some simplifying assumptions. The N-M theory is no exception. For one thing, N-M assume that the players can communicate freely; that is, communicate and/or act together as they please. Ideally, all the players can communicate simultaneously; in practice, of course, they can't. And the deviation that occurs in practice from the ideal is very important. It has been shown experimentally that the physical arrangement of the players affects the bargaining, and players who are aggressive and quick to make an impact do better than others who are more reticent.

In addition, N-M assume that utilities are freely transferable between the players. If the payoff is in dollars, for instance, and *A* pays *B* one dollar, *A*'s gain in utiles is the same as *B*'s loss in utiles. This is a severe restriction on the theory and probably its weakest link. Actually, the assumption is not so restrictive as it may seem: it *does not* require that *A*'s pain at losing a dollar be the same as *B*'s pleasure in getting it. Also, there are many ways of assigning utilities to each of the players. However, if *n* people are in a game, it is unlikely that appropriate choices of utility functions could be made that would satisfy the restriction.

Vickrey's Self-Policing Imputation Sets

N-M's initial work on the n-person game aroused a great deal of interest. The theoretical framework which they fashioned was adopted, at least to some extent, in subsequent work. William Vickrey, for example, assumed that the n-person game was described in characteristic function form and that the final outcome would be an imputation, just as N-M had done. But although Vickrey was concerned with imputation sets which reflect "standards of behavior," he did not restrict himself to the study of N-M solutions but classified arbitrary imputation sets on the basis of their "stability." He tried to determine "what set of imputations, if adopted as approved standards or patterns of distribution, stands the best chance of being adhered to, or alternately, what sets of imputations require sanctions of minimum strength in order to compel adherence."

Specifically, Vickrey's argument went something like this: Suppose there is some imputation set which reflects the accepted standard of behavior, and suppose further that the present state of society is one of the imputations represented within the set. Consider what happens when that imputation is replaced by another which lies outside the accepted imputation set and which dominates it with effective set H. The new imputation is not in accordance with the accepted standard of behavior, so the set H is called the *heretical set,* and the imputation itself, a *heretical imputation.*

This heretical imputation can only be in effect temporarily. Eventually, social disapproval or whatever sanctions enforce society's standards will induce a change. The heretical imputation will be replaced by another which dominates it and which, once again, lies in the original imputation set. In fact, society's pressures may be so strong that no heresy can occur at all. "If the social . . . disapproval [is] strong enough, and particularly if [it] becomes entrenched in formal rules of law, then adherence to any given set of imputations can be enforced and all others proscribed. Indeed, the proscription of certain imputations may have as much force as the rules of the game from which the characteristic function form is determined: but if this is the case then it would seem hardly worth the trouble to work through the characteristic function when these social norms could be appealed to directly."

What we are dealing with here is a standard which is not so rigidly enforced that it can't be violated. In the long run, however, societal pressures are strong enough to prevent the survival of heresies. The key to determining stability, therefore, lies in the heretical set H. By comparing the fortunes of the players of H in the original and the final imputations, one can anticipate whether the heresy is likely to occur. If the players in the heretical sets wind up with less than they would originally, the original imputation set is considered stable, since, in order to effect the heresy, the consent of all the players of H is required, and it is unlikely to be obtained in this case. If the players do better in the final imputation than they would in the original one, the original imputation set is considered unstable.

An imputation set is not simply stable or unstable. There are, rather, degrees of stability. In one instance, a heresy will inevitably be followed by a lower payoff for *one particular* player. In another instance, a heresy will be followed by a lower payoff for *some* player, but who that player is will not be specified; it will depend on the imputation to which society finally returns. Obviously, the last instance is less stable. Although each player knows someone will have to suffer, they can each hope it will be someone else. In the first instance, there is no room for hope for the unfortunate player. To consummate the heresy, the heretical set *H* will need his approval, and they are not likely to get it. Both these imputation sets are more stable, however, than one in which none of the heretics is punished for his transgression.

Self-Policing Patterns

Suppose a set of imputations—which we will call a *pattern* —represents a standard of behavior, and among these imputations, one, *x*, represents the status quo. An imputation *y* which dominates *x* and which lies outside the original pattern is called a heretical imputation, and the players in *H*, the effective set, are called heretics. Now consider all the imputations *z* which lie inside the pattern and which dominate *y*. These are called *policing imputations* and each of them is potentially the imputation to which society will eventually return. For each heretical imputation *y*, there are usually many policing imputations *z*.

Now let's fix our attention on one particular heresy *y* and all the imputations *z* which dominate *y* and lie in the original pattern. If there is a player *i* in *H* who receives less in *every z* than he obtains in *x*, the heresy *y* is called *suicidal* for player *i*. Once the heresy *y* is formed, *i* knows he will inevitably lose.

Now let us go back to *x*, the original imputation in the

pattern. We say that x is a *strong imputation* if every heretical imputation y which dominates x with effective set H is suicidal for some player in H. It is to be expected that different players will be punished for different heresies. What is important is that every *particular* heresy have some *particular* player's name on it—the player who will inevitably suffer, and who presumably will break up the coalition. It doesn't matter if the name of the suicidal player changes as you pass from one heretical imputation to another.

Finally, if every imputation in a pattern is called strong relative to that pattern, we say that the pattern is *self-policing*.

Let us look at some of these ideas in the context of a particular example. Suppose a game we talked about earlier is still being played—the three-person game in which a player alone gets nothing but any coalition of two or three players can make 2. Suppose the pattern being considered is the symmetric solution in the N-M sense: the set of imputations $(0, 1, 1)$, $(1, 0, 1)$, and $(1, 1, 0)$. And, finally, suppose the original imputation x is $(0, 1, 1)$.

It is immediately clear that the heretic set H must include player A; B and C are already getting their limit in the original payoff. For the sake of definiteness, suppose the set H consists of A and B. This means that, in the heretical imputation y, B must get more than 1, since he obtains 1 in x. It also means that, when the policing imputation z is formed, B can't be in the effective set. This follows from the fact that z dominates y, z is in the original pattern, and in the original pattern no payoff is greater than 1. So z must be specifically $(1, 0, 1)$. Therefore, y is suicidal for B. An obvious, symmetrical argument could be applied if the heretical coalition consisted of A and C or if the original imputation x were either of the other two imputations in the pattern. We conclude that the pattern is self-policing (see Figure 45).

Now look at the discriminatory solution: all imputations in which player A gets $1/2$. Suppose the original imputation

x is $(1/2, 1\ 1/2, 0)$ and the heretical imputation y is $(1, 0, 1)$, with H consisting of A and C as the effective set. The policing imputation z must be in the solution, which means that A must get $1/2$ in z. Since A is getting more in y than

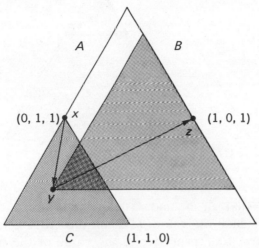

Figure 45

he would get in z, he cannot be in the effective set of z, and therefore the effective set of z must be B and C. So C must get more than 1 in z. C not only gains in the short run (going from x to y); he gains in the long run as well (going from x to z). A, the other member of the heretical set, doesn't lose either. For a while he's ahead, and then he goes back to what he started with. The only one to lose is B, the one player who was not in H. The discriminatory solution is clearly not self-policing (see Figure 46).

The Aumann–Maschler Theory of n-Person Games

The Aumann–Maschler theory (which we will denote A-M) is similar to the N-M theory in that it also uses the characteristic function form description of the n-person game, but in almost all other respects it is quite different.

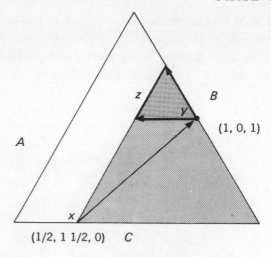

Figure 46

Suppose that A, B, and C are players in a three-person game in which the values of coalitions AB, AC, and BC are 60, 80, and 100, respectively; the value of the three-person coalition is 105; and the value of every one-person coalition is zero. The game is illustrated in the diagram in Figure 47.

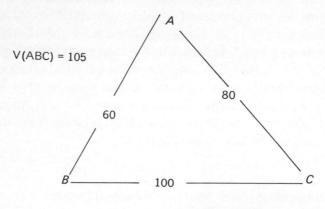

Figure 47

The A-M theory does not attempt to predict which coalition will form; its purpose is to determine what the payoffs would be once a coalition is formed. The theory takes only the strengths of the players into account; all considerations of fair play and equity are put aside. Before explaining precisely what this means, let us assume that players A and B have tentatively agreed to form a coalition and are now concerned with dividing their payoff of 60. Their conversation might go something like this:

Player B (to player A): I would like my payoff to be 45, and so I offer you 15. I have a stronger position than you do in this game, and I think the payoffs should reflect it.

Player A: I refuse. I can certainly get 40 by going to player C and splitting the coalition value evenly with him. He would accept my offer, since at the moment he is getting nothing. But I'm not greedy. I'll accept an even split with you.

Player B: You're being unreasonable. No matter what I offer you, even if I offer you the whole 60, you can always threaten to go to C and get more. But your threat is based on an illusion. At the moment C is in danger of getting nothing, so he's receptive to any offer that comes his way. But remember that defecting to C is a game I can play too. Once our tentative coalition disintegrates, you won't have C to yourself. You'll have to compete with me, and I have the advantage. If I join C our coalition would be worth 100, whereas yours would be worth only 80. If your objection to the 45–15 split were accepted as valid, then no agreement between us would be possible. If it ever comes to a fight between us for a partnership with C, we would each stand a chance of being left out entirely. I think we'd both be better off if you accept my reasonable proposal.

At this point, let us leave the bargaining table and take a closer look at the players' arguments. B's initial proposal seems unfair, since it prescribes that A get a much smaller share than B, and from one point of view it should be ruled out. But in the A-M theory the element of equity is not taken into account. If B's proposal is to be eliminated as a possible outcome, it must be on other grounds. (If a payoff is not precluded, that does not mean that it is *the* predicted outcome. A-M, like N-M, recognize that there may be many possible payoffs.)

A's objection—that he can go to C and offer C more than he would get from B (which is nothing) and still have more left over for himself than B offers him—is more pertinent. In effect, A is saying that if he and C have the power to form a new coalition in which they both get more than B would give them, there must be something wrong with B's proposal. Specifically, A feels he isn't getting enough.

In his answer, B puts his finger on the flaw in A's argument, which is that it goes too far. If A's argument is allowed to rule out B's proposal, it must rule out every other proposal as well. Even if B yielded the entire 60 to A, he would be open to the same objection, since A could still offer C 10 and receive 70 for himself. And, for that matter, so could B. If *any* coalition is to be formed, it must be in the face of this objection.

Granted, then, that A's objection is not valid, and granted that B has a competitive advantage with C, why should B get 45? Why not 50? Or 40? A-M take the view that B should not get 45, and the reasoning goes something like this.

Suppose A persists in demanding more than 15 and B refuses to yield. Then A and B will have to compete for C as a partner. Presumably, B would demand 45 from C also, for he wants 45 and it's all the same to him where he gets it. But if B gets 45 in the coalition BC, that leaves 55 for C. A can offer C 60 and take 20 for himself, which is 5 more than he was offered by B. Of course, B can always lower his demands, but then why ask so much from A to begin with? It turns out that the "proper" payoff when coalition AB forms is 40 for B and 20 for A.

As another illustration of this kind of argument, consider what happens when a three-person coalition is formed. A-M maintain that the payoffs to A, B, and C should be 15, 35, and 55, respectively. To see why, suppose we start with some other payoff—say 20, 35, 50—and see what goes wrong.

Since C is the one who seems to have been slighted, he

should be the one to object, and so he does. He offers *B* 45, takes 55 for himself, and *A* has no recourse. There is only one way *A* can match *C*'s offer: he must also offer *B* 45, and to do this, he has to lower his original demand of 20 to 15. A-M interpret this to mean that *A* was too ambitious in the first place. Of course, the payoff suggested by A-M, (15, 35, 55) is open to the same objection. *A* might, for instance, offer *B* 37 and take 23 for himself. But now, without raising his original demand of 55, *C* can outbid *A* by offering *B* 45.

This discussion gives a rough indication of the considerations that underlie the A-M theory. The next step is to reformulate in a more precise way what we have just said.

When the A-M theory of n-person games was first conceived, it was hoped that it would enable one to determine which coalitions would form and which payoffs would be appropriate. For tactical reasons, the first problem was postponed and A-M addressed themselves to the second: Given a particular coalition structure, what should the appropriate payoffs be? It was anticipated that once a payoff could be associated with each potential coalition formation, it would be possible to predict which coalition would form. So far, however, nothing has been done along this line. The fact is that you often can't predict which coalitions will form, even when you know (or think you know) how each coalition would distribute its payoffs. To see why, let's go back to our last example, in which the two-person coalitions had values of 60, 80, and 100. In this game, according to A-M, the payoff of a player is independent of *which* coalition he enters. The only thing that matters is that he joins a coalition. That is, *A* receives 20, *B* receives 40, and *C* receives 60 if they enter *any* two-person coalition. (If they form a three-person coalition, they each get 5 less.) Consequently, each person is indifferent whom he joins, as long as he joins someone, and it is impossible to say, on the basis of the formal rules alone, that any two-person coalition is more likely to form than another.

The Formal Structure

A-M start by assuming the existence of a coalition structure in which each player is in precisely one coalition—possibly a one-person coalition consisting of only the player himself. (They do not go into the question of whether the coalition is advisable or likely to form.) They assume (with no loss of generality) that all one-person coalitions have a zero value. Then a tentative payoff is assigned to each of the players, subject to certain restraints. The sum of the payoffs in any coalition should equal the value of that coalition; the coalition doesn't have the power to get any more, but it can see to it that it doesn't get any less. In addition, no player ever receives a negative payoff, for if he did, the deprived player would prefer to stand alone.

With these preliminaries out of the way, let's go back to the basic question: When is the payoff appropriate for the given coalition structure? In order to answer this question, A-M first look for the answer to another, subsidiary question: Is there any player with a "valid" *objection* against another player in his own coalition?

The Objection

An objection of player i against player j is simply a proposal that a new coalition be formed, call it S, with certain payoffs. Both players must be in the same coalition in the original structure. Otherwise, i would have no claim over j. If the objection is to be valid, player i must be in S but player j must not; each player in S must get more than he was getting originally; and the sum of the payoffs of all the players in S must equal $V(S)$. The reason for all these conditions is clear: if the objection serves to show j that i can do better without him, i can hardly ask for j's cooperation in forming a new coalition. Also, the players in S must be paid more, or they'd have no motivation to join. Finally, $V(S)$ is all that S can be sure of getting, and there's no reason why they should settle for anything less. In effect, i's objection to j is the same that A made to B in our earlier example: "I can do better by joining S. The other prospective members of S also do better, so I'll have no trouble convincing them to join me. If I don't get a bigger share in the present coalition, I'll go elsewhere."

If, for a given payoff, no player has a valid objection against any other, A-M consider the payoff acceptable for that coalition structure. But what of the converse? If a player does have a valid objection against another, should it necessarily rule out the payoff? A-M think not. Otherwise, in some games every payoff in every coalition structure would be ruled out (except where only one-person coalitions are formed and objections are impossible). As a matter of fact, our original example was just such a game. So that there can be *some* acceptable payoffs, A-M, under certain conditions, will allow the original payoffs to stand if j can truthfully counter-object: "I too can do better by leaving the coalition and joining some of the other players."

At first glance, j's reply does not seem to be really responsive to i's objection. After all, if they both can do better, perhaps they both should do better. The answer of j seems to confirm what i suggested: the original structure should be broken up. There are two difficulties, however. In the first place, the "others" that i and j are threatening to join may be the same player(s). (In our example, he was.) And so i and j can't do better simultaneously. Also, when the original coalition dissolves, i and j are forced to be dependent on other players, so they have not necessarily improved their position. A-M assume that the members of a coalition would prefer to keep control over their own fate and will look elsewhere for partners only if some player persists in making demands that are disproportionate to his power. If a player can do better by defecting and his partner cannot, A-M feel that the second player is asking too much. But if *both* players can do better by defecting, it may behoove them both to sit still rather than go round and round and risk being left out of the final coalition. A-M consider a payoff to be acceptable, or *stable*, if, whenever a player i has a valid objection to j, j has a *counter-objection* to i.

The Counter-Objection

A *counter-objection*, like an objection, is a proposal for a coalition T, with a corresponding payoff. It is made by a player j in response to an objection directed toward him by another player i. T must contain j, but not i. The point of the counter-objection is to convince the objector that he is not the only one who can do better by defecting, and it is essential that j make it plausible that such a coalition can form. To induce the players in T to join him, he must offer them at least what i offered them, if they happen to be in the objecting set S, and if not, what they would have obtained originally. But the sum of the payoffs must not exceed what he can afford to pay, $V(T)$, the value of the new coalition. Finally, j must receive at least as much as he would have obtained originally.

The set of stable payoffs for a given coalition structure—which we will call the A-M solution—is called the A-M *bargaining set*.

A few points in the A-M theory should be stressed. For one thing, stable payoffs need not be fair. A-M are not seeking equitable outcomes but outcomes which are in some sense enforceable. Suppose, for instance, that the two-person coalitions have values of 60, 80, and 100, just as before, but the three-person coalition has a value of 1,000 rather than 105. Suppose further that a three-person coalition forms with a payoff of (700, 200, 100). Judging from the values of the two-person coalitions, C seems to be stronger than B, and *B* seems to be stronger than *A*. Nevertheless, A-M consider this payoff—in which the "weakest" player gets the most and the "strongest" player the least—stable. The reason for this is that *B* and *C*, despite their complaints, have no recourse. They can break up the coalition, of course, but then they would lose too, along with *A*. It is true that *A* would lose more, but the A-M theory does not recognize dog-in-the-manger tactics. For a player's complaints to be considered valid, he must be able to do *better* elsewhere. Thus, A-M avoid making interpersonal comparisons of utility, such as comparing a loss of 700 by *A* with a loss of 100 by *C*.

The Aumann–Maschler and von Neumann–Morgenstern Theories: A Comparison

The most important difference between the A-M and N-M theories is in their concept of a "solution." For N-M, the basic unit is a set of imputations. An imputation, taken alone, is neither acceptable nor unacceptable but can be judged only in conjunction with other imputations. In the A-M theory, an outcome stands or falls on its own merits.

In the last sentence we referred to an outcome rather than an imputation, and this is another difference between the two theories: A-M do not assume the outcome will be Pareto optimal. On the face of it, it may seem absurd for players to decide on one outcome when with another outcome they all do better. Absurd or not, however, this is what often happens. Moreover, non-Pareto-optimal outcomes are important in the A-M theory. A-M are very much concerned with determining which coalitions and payoffs will hold under the pressure of players trying to improve their payoffs. The only weapon that players have is a threat of defection, and if this is to be plausible, defections must occasionally occur. Often the new coalitions are not Pareto optimal. Workers must strike occasionally, then, if threats are to be taken seriously, and when they do, the outcome is generally not Pareto optimal.

Perhaps the biggest advantage of the A-M theory is that it does not require interpersonal comparisons of utility. The payoffs may be given in dollars without affecting the theory; all that is required is that a person prefer a larger to a smaller amount of money. The A-M theory also drops the assumption of superadditivity, but this is less important. It is still a reasonable assumption but is no longer needed.

The A-M solution, like the N-M solution, sometimes consists of more than one payoff (for a given coalition structure), and the theory makes no attempt to discriminate between them. In our example in which the three-person coalition received 1,000, there are many stable payoffs for the three-person coalition. Threats play a very limited role —they preclude only the most extreme payoffs—and the game is one of almost pure bargaining.

For a given game, there are generally many outcomes which are consistent with both the A-M and the N-M theories. But, in a sense, much greater variety of N-M solutions is possible, and this makes the N-M theory much more comprehensive. In the three-person game, for example, in which any coalition with more than one player received 2 and single players received nothing, each coalition structure had only one stable payoff. The N-M symmetric solution for this same game consists of the three stable A-M imputations, and there is nothing in the A-M theory which corresponds to the N-M discriminatory solution. The greater variety of solutions in the N-M theory leads to greater flexibility but makes the theory almost impossible to test in practice. The A-M theory at times seems to narrow but is much easier to test. Consider the following example:

An employer wishes to hire one or two potential employees, A and B. The employer together with the employee will make a combined profit of $100. Together the employees get nothing, and the employer by himself gets nothing. All three may unite, and then the profit will also be $100.

In the A-M theory there are, basically, two possible outcomes: either the employer forms a coalition (with one or both employees) and gets $100, or he doesn't and gets nothing. The workers get nothing in any case. If one worker offered to take $10 and give the employer $90, the employer could counter by offering the other worker only $5 and the first worker would have no recourse. Thus, any payoff in which a worker receives anything is unstable.

A-M solutions are very convincing in certain situations. When there are a large number of workers who cannot communicate easily, they may very well compete in the way suggested by the A-M theory. But at other times it doesn't seem to apply. It is fairly obvious that unbridled competition hurts all the workers, and not surprisingly, in real life workers often join together to act as a single coalition even though they gain nothing immediately but bargaining power. In effect, the game is reduced to a two-person bargaining game, with the workers acting as a single player. (In large industries, such as steel or motors, companies generally do not use wage levels to compete for the best worker; nor do workers offer to take a cut in salary to get a job. Both industry and labor merge into what Galbraith calls countervailing forces, and the result is in effect a two-person game.)

The N-M theory, on the other hand, has many solutions for this game. One solution consists of all imputations in which each worker gets the same amount. This may be interpreted as follows: the two workers join and agree to split anything they get equally, and they negotiate with the employer for $100. This results in a two-person bargaining game in which anything can happen. It is interesting that the sole payoff dictated by the A-M theory—$100 for the employer and nothing for the workers (assuming *some* coalition forms)—is contained in every N-M solution.

One last illustration of A-M's concern with enforceable rather than equitable outcomes is given below.

Each of two retailers A and B has a customer for a certain item who is willing to pay $20, and each of two wholesalers C and D has a source at which he can obtain the item for $10. In the four-person game consisting of A, B, C, and D, the value of a four-person coalition is $20; the value of any three-person coalition is $10; the value of the two-person coalitions AC, AD, BC, BD is each $10. The value of every other coalition is zero.

Now consider all the coalition structures in which the players get $20, the greatest amount possible. There are three possibilities: either of two two-person coalitions: (AD and BC) or (AC and BD); or the four-person coalition. In each of the three instances, both wholesalers will get the same payoff, and both retailers will get the same payoff; but, in general, the wholesalers will get different payoffs from the retailers. This seems odd when you consider that the wholesalers and retailers have completely symmetric roles. But A-M's reasoning is this: If A is getting less than B, he can join the wholesaler who is getting the lowest payoff and they will both do better. If both retailers and both wholesalers are getting the same amount, there is no way to form a new coalition in which all the players do better. (For the sake of simplicity, we have taken some liberties with the A-M theory as it was originally formulated. We omitted certain assumptions about coalitional rationality, for example, and restricted the size of the objecting and counter-objecting sets to a single player. This introduces some changes—for one thing, it assures at least one stable payoff for every coalition structure—but the spirit remains the same.)

The Shapley Value

Shapley approached the n-person game in still another way. He looked at the game from the point of view of the players and tried to answer the question: Given the characteristic function of a game, of what value is that game to a particular player?

From what we have seen, predicting the outcome of an arbitrary n-person game on the basis of the characteristic function alone would seem to be a hazardous business. The personality of the players, their physical arrangement, social custom, communication facilities, all have some effect on the final payoff. Nevertheless, Shapley found a method for calculating the worth of a game to each player—generally called the Shapley value—on the basis of the characteristic function alone. This number was arrived at a priori, with all other relevant factors abstracted away.

Shapley's scheme is only one of many that could serve the purpose. Why use one and not another? Shapley justifies his choice as follows: He lists three requirements that he feels any reasonable scheme should satisfy, and then he goes on to show that his scheme satisfies these axioms, indeed, that his scheme is the *only* one that does. The critical requirements are these:

1. *The value of a game to a player depends only on the characteristic function.* This means the values are assigned without regard to the identities or traits of the players. In a bargaining game, for example, in which the two players get nothing when they stand alone but share something when they get together, their values would be the same.

2. *A payoff in which each player receives his value is an imputation.* Shapley assumes that rational players will form an imputation. (He also accepts N-M's assumptions of superadditivity and the transferability of utilities.) Since the sum of the payoffs is necessarily equal to the value of the n-person coalition (by the definition of imputation), and the value of the game to a player is his average payoff (in some sense), it follows that the sum of all the values should also equal the value of the n-person coalition.

3. *The value to a player of a composite game is equal to the sum of the values of the component games.* Suppose a group of players is simultaneously engaged in two games.

Define a new game with these same players in which the value of a coalition is equal to the sum of the values that it had in the original two games. In this new game, each player has a Shapley value, and axiom 3 states that it should equal the sum of the players' Shapley values in the original

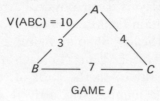

GAME *I*

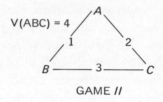

GAME *II*

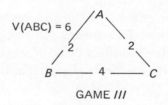

GAME *III*

Figure 48

two games. Figure 48 may make this clearer. Note that game *I* is a composite of games *II* and *III*. Coalition *BC*, which has a value of 3 in game *II* and a value of 4 in game *III*, for example, must have a value of 7 in game *I*. Axiom 3 states that in such a situation the value of game *I* to a player must be the sum of the values of game *II* and game *III*.

For a more critical discussion of Shapley values the reader should turn to *Games and Decisions* by Luce and Raiffa. Two games which we referred to earlier are shown in Figure 49, with the Shapley values of each of the players.

For the mathematically sophisticated, this is the way one

computes the Shapley value $V(i)$, for an arbitrary player i, in an arbitrary n-person game:

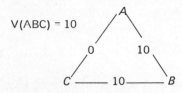

V(ABC) = 10

The Shapley value for:
Player *A* is 5/3
Player *B* is 20/3
Player *C* is 5/3

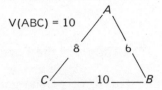

V(ABC) = 10

The Shapley value for:
Player *A* is 7/3
Player *B* is 10/3
Player *C* is 13/3

Figure 49

For each coalition S, let $D(S)$ be the difference between the values of the coalition S and the coalition S without player i (if i is not in S, $D(S) = 0$). For each coalition S, *compute* $[(s\text{-}1)!(n\text{-}s)!/n!] \cdot D(S)$, where s is the number of players in S, n is the number of players in the game, and $n!$ means $n(n\text{-}1) \ldots (3)(2)(1)$. Add these numbers for all coalitions S, and the answer is the Shapley value for i.

Shapley's formula may also be derived from a bargaining model. Imagine that at the start one player joins another to form an intermediate two-person coalition, then these two are joined by a third player, and ultimately an n-person coalition is formed, one player at a time. Suppose that at each stage the new player gets the marginal gain: the difference in the values of the coalition already formed and the coalition with the new player. If one assumes that the final n-person coalition is as likely to form in one way as in another, the expected gain of a player is his Shapley value.

This derivation of the Shapley value is interesting but not very convincing. The Shapley value really stands or falls on the basis of the axioms. There is some experimental evidence to the effect that coalitions actually are formed one player at a time. Even so, it is unreasonable to expect the new player to get all the marginal gain. And it is not clear that the orders of formation are all equally likely.

The Theory of the Triad

The three-person game, which is just a special case of the n-person game, is particularly interesting because it is so simple. Relationships that are obscured in larger, more complex games are often clearer in the three-person game. Most of our examples are of three-person games for just this reason.

Social scientists who have been concerned with blocs and power have also generally used the three-person game as a model. Theodore Caplow, for example, was primarily concerned with the way coalitions are formed and the way payoffs are distributed among players with varying amounts of power. He constructed a theory which he then applied to the three-person, weighted-majority game. (In a weighted-majority game, each player has a certain weight. A coalition has a value of 1 if the sum of the weights of the players in that coalition is more than half the total weight of all the players. Otherwise, the coalition value is 0. The weight of a player is sometimes called his "strength.")

Caplow feels that a number of factors determine how coalitions are formed. One factor is the frequency with which the game is played; that is, whether it is played continuously, periodically, or only once. Even more important, in his opinion, are the relative strengths of the players, each of whom is trying to gain power or control over the others.

Caplow studied six different types of games in which power is distributed among the players in different ways. These power distributions can be described by three numbers, (3, 2, 2), for example, signifying that the first player has a weight of 3 and the other two players a weight of 2. In qualitative terms, players *II* and *III* have the same strength, each is weaker than player *I*, and together they are stronger than player *I*. The six games that Caplow studied had strength distributions $(1, 1, 1)$, $(3, 2, 2)$, $(1, 2, 2)$, $(3, 1, 1)$, $(4, 3, 2)$, and $(4, 2, 1)$. In each game, Caplow predicted whether a two-person coalition would form, and which (if any) it would be.

According to Caplow, the coalition that eventually forms must contain half of the total strength. But since there is necessarily more than one such coalition, this qualification is not definitive. Of the coalitions that control a majority of the power, Caplow feels, the one with the *least* total power would be the one to form. After the coalition forms, the payoffs would reflect the strengths of the players.

If Caplow is right, no three-person coalition could ever form, since the two strongest players always have a majority of the strength. Indeed, there are just two possibilities: either the strongest player has sufficient strength to constitute a majority coalition by himself, or the two weakest players will form a majority coalition. (We disregard the special case in which the strength of the strongest player is exactly half the total strength.)

It may seem odd that the coalition with the least, rather than the most, total power is the one that will form, but the reasoning is simple enough. If a player can form a majority coalition with one of two other players, he should—if you accept Caplow's premises—prefer the weaker one. The coalition will have the preponderance of strength in any case, and the weaker player will demand a smaller share of the profits. In the game $(4, 3, 2)$, player *I* will demand the lion's share in any coalition he forms. Player *III* would prefer to join player *II* and, conversely, player *II* would prefer to join player *III*. Of course, player *I* would also rather join player *III* than player *II*, but his preferences are academic since it is *II* and *III* who will eventually join.

Several attempts have been made to test this theory experimentally. Generally, the experiments were designed along the following lines:

Each player was assigned a certain weight. Three markers, one for each player, were placed on the first square of a board consisting of a number of consecutive squares. Both markers and board were much like those used in the parlor game Monopoly. A die was thrown, a number came up, and each player moved his marker forward x squares, x being the product of the player's weight and the number on the die. The player who finished first won. If this were all, the player with the greatest weight would always win— but one variation was permitted: two players might join together and form a coalition. When this happened, each member of the coalition moved forward a number of squares equal to the product of the face of the die and the *sum* of both players' weights. If the die showed 3 and the players' weights were 4 and 6, each player would move forward thirty squares. If a coalition was formed, the players must win or lose together, since they always moved forward the same number of squares. At the time of formation, each coalition announced how it would distribute its profits if it won.

In the actual experiments, it was soon evident to the players that, once a coalition was formed, the outcome was determined. There was no need to play the game through to the end. It was also true, if not quite as obvious, *that the weights were only window dressing*. To see why, re-examine the game in which the weights are distributed $(4, 3, 2)$. If a two-person coalition forms—any two-person coalition—it will win and the third player will lose. Each of the players faces an identical problem: getting into the coalition. In effect, all the players have the same strength, and the distributions $(4, 3, 2)$, $(1, 1, 1)$, $(3, 2, 2)$, and $(1, 2, 2)$ constitute the same game as do $(3, 1, 1)$ and $(4, 2, 1)$. (Once again, this presupposes that *some* majority coalition will form.)

The Experimental Evidence

Although the weights have little significance in themselves —they are meaningful only insofar as they indicate whether any one player has a majority of the strength—they had a strong influence on the play. There is a simple explanation for this paradox: the players *believed* the weights were significant. Often the "strong" players, those with a plurality but not a majority of the weight, would demand a larger share in any coalition they might join, and when they did, the other players stayed away just as predicted. Even when the other players acknowledged that the larger demands were legitimate, the effect was the same: the "strong" player was priced out of the market. The illusory extra "strength" worked to the player's detriment.

In one experiment, the Caplow hypothesis was tested and generally confirmed. However, in a subsequent experiment some doubts were raised. It was felt that the players misperceived the game because of its complexity, that six different games were too many to cope with. In a second experiment, then, only one weight distribution was used, (4, 3, 2), and this game was played repeatedly.

At first, the strong players were excluded, as they had been before, though the players thought the extra demands were justified. Initially, seventy-six of ninety subjects indicated that it was proper for a player with a weight of 4 to ask for more than half in any coalition he entered. At the end of the trials, however, virtually every player realized that the weights made no difference. The "strong" players asked no more than the others and were no longer excluded from coalitions.

In another experiment, the "strengths" of the players were defined in a somewhat different way. At the start, each player was assigned a quota: (0, 2, 4) for players *I*, *II*, and *III*, respectively. Then each player decided (independently of the others) which of the other two he would most like to join in a coalition. The pair of players that chose each other —if two did—was given a minute to decide how to divide 10 units. If they reached an agreement, the players in the coalition received the amount agreed upon and the third player obtained his quota. If they failed to agree, each player received his quota.

In this game, the quotas *were* significant and were properly taken into account. When the subjects started to play, it was generally agreed that player *III*, with a quota of 4, had the advantage. This was still the opinion of most players when the game ended, though some then felt that player *II*, with a weight of 2, had the advantage.

(According to the A-M theory, all the players should demand 5 if they enter into a coalition. If they did, there would be no reason for one coalition to form rather than another. It wouldn't follow from this that all players get the same expected payoff, however, since they get varying amounts when they are excluded from the coalition. If we assume that one two-person coalition is as likely to form as another, and that one two-person coalition always forms, the average payoff for players *I*, *II* and *III* would be [10/3, 12/3, 14/3], respectively.)

In another experiment, which took the form of the round-the-board game we mentioned earlier, three different weight distributions were used: (6, 6, 1), (6, 6, 3), and (6, 6, 5). In this game, if no two-person coalition was formed, the game was repeated. The point of the experiment was to test the following hypotheses:

1. Weak-strong coalitions form more often than chance would suggest.

2. If a strong player joins a weak one, he will get more than the weak player.

3. The weak player in the (6, 6, 5) game will enter fewer coalitions but will obtain a larger share in the coalitions he enters than the weak players in the (6, 6, 1) or (6, 6, 3) games.

The first hypothesis proved to be incorrect. The number

of weak-strong coalitions was no more or less than one would expect if the coalitions were formed at random. (And this is exactly how it should be if the players are sensible.) Oddly enough, the second hypothesis seemed to be correct. The strong players did get more when they joined weak ones. The third hypothesis proved to be partially true and partially false. It was not true that weak players entered fewer coalitions when their weights increased, but they did seem to get larger payoffs. (On this point, however, the experiment was not conclusive.)

There is an interesting paradox here. Suppose that in the (2, 2, 3) game the subjects act as though the weights are significant and, as Caplow suggested, exclude player *III*. Could the players be said to be acting irrationally? It would seem so, and yet it is impossible to single out any one player who is acting against his own interests. The two weak players can do no better than join together. Even if the strong player gives up any claim and joins the others on an equal footing, the weak players would be no better off. They would be in a winning coalition two-thirds of the time, and as it is, they are always in a winning coalition. The strong player may not be happy about this, but he can hardly be called irrational, since there's nothing he can do about it. Conceivably, the weak players might take their lesser weights as a cue to get together as Schelling suggests, but, in any case, it is difficult to see how one could criticize their actions.

Interestingly enough, players who score as "high achievers" on psychological tests and who presumably are eager to do well initiated more offers than the others, whether they had weak or strong positions, and accepted smaller portions in order to enter winning coalitions. They were also more sensitive to incipient coalitions being formed by others. The social class of the subjects, on the other hand, had almost no effect on performance. Women were consistently less competitive than men. Women tended to form three-person coalitions whenever possible and were reluctant to exclude anyone, even if this meant they themselves were ultimately excluded. To women, equity was more important than winning; one woman went so far as to instigate two other players to join against her. Men were highly competitive, and this often worked to their detriment. Because their demands were greater, they were excluded from coalitions more often.

Voting Games

A group of individuals may at times be called upon to make a single, joint decision. Families, legislative bodies, stockholders, national electorates, committees, juries, the Supreme Court of the United States, are all in this position. The decision of the group is dependent upon the decisions of the membership, but not always in the same way. Everyone may be allowed a single vote; or the votes may be weighted to reflect the wealth of the voters or the number of shares owned by each one. There may be an absolute veto such as there is, under certain conditions, at the United Nations, or a partial veto such as is available to the President of the United States or to the House of Lords in England.

The normal form is well suited to describe this kind of voting game. The strategies are the choices listed on the ballot; these are clear cut and known to all the players. The voting procedure that determines the payoffs (who wins the election or what bill passes) is also known to the players. There is some variation in the other variables, however. In a national election, for instance, some communication exists between one voter and another and between the candidates and the electorate as a whole, but this kind of communication is of a very different order from the "logrolling" that takes place in a legislative body. What makes the study of these games most attractive to the game-theorist is the availability of a large number of ready-made "laboratories" in which the parties are playing for keeps.

On the face of it, voting games appear to be very simple: The players cast their votes for the choice they find most attractive. But, in fact, these games can be highly complicated. To see what kinds of problems may arise, let us consider an example from *The Calculus of Consent* by James Buchanan and Gordon Tullock.

There are a hundred farms in a township. Each farmer has his own local road that leads to the main highway. Each of these local roads is maintained by the township, and when a road falls into disrepair, all the farmers vote on whether to fix it. Even in this simple model there are several possible outcomes:

1. *Everyone votes on principle.* Each farmer sets a uniform standard and uses it to determine when his own and his neighbors' roads should be repaired. In general, the standard will vary from farmer to farmer. In effect, each farmer decides on an ideal state of repair in which the advantage of having his own road in a good state of repair is balanced against the cost of maintaining everyone else's road in the same state of repair. On the basis of this, he votes for repairing all below standard roads and against repairing the others. In effect, a road will be repaired when it falls below the median standard.

2. *A majority votes on principle.* Most of the farmers vote on principle, but a few vote to repair only their own roads. As a result, the quality of the heretics' roads is raised, and everyone else's is lowered.

3. *Cutthroat voting.* Everybody plays the heretics' game and no one's road is repaired.

4. *Bloc voting.* A majority of the farmers, say 51 percent, join together and form a coalition. This group sets a standard of road repair for its *own* members and maintains it by voting as a bloc. No other road repair is ever approved, and outsiders are helpless. The members are much better off than they were in situation 1, since they pay only 51 percent of the money allocated to road repair, yet receive all the benefits.

Condition 2 is obviously unstable. If the electorate is alert and can communicate, and if the game is played repeatedly, condition 2 won't obtain very long. Either the game will degenerate into situation 3, or a coalition will force the game into situation 4. (It is even conceivable that a majority within the majority may take control in situation 4, but if it rules with too heavy a hand, it jeopardizes the larger coalition.) In the absence of communication, the game is like an n-person "prisoner's dilemma" in which situation 1 is the cooperative outcome and situation 3 is the noncooperative outcome.

Whatever plausibility these outcomes have as realistic solutions for our voting model, two of them are grossly unfair. In both 2 and 4, the farmers play identical roles, yet some of them fare considerably better than others. This suggests a more general question: Is there an overall way to translate the individual preferences of the electorate into a group decision that will be equitable (in some sense) and related to the individual preferences?

Suppose an arbitrator was given the list of possible choices and was asked to select one on the basis of the players' preferences (the order of the preferences, that is, the intensities of the preferences, is disregarded). What is wanted is a well-defined numerical technique and not some qualitative rule such as Jeremy Bentham's "greatest happiness for the greatest number."

One possibility is to consider the choices two at a time in some arbitrary order and for each pair to determine whether a majority of the electorate prefer one or the other. The preferred choice is then compared with another, and so on, until the list is exhausted. The choice that remains at the end is the winning one. The difficulty here is that the order in which the choices are considered may determine which one wins, and this order is completely arbitrary. This was the case in the school-construction example that we discussed earlier. There, it will be recalled, no one plan was preferred to all other plans.

The arbitrator's problem, then, is not only difficult but in a sense may even be impossible. Suppose that the arbitrator is told the relative preferences of each member of a group with respect to certain choices, as before. (We assume that there are at least two members in the group and at least three choices.) Suppose, further, that the arbitrator must derive an order of preference for the group, based on the order of preference of the individuals in the group. In other words, he must come up with a scheme that translates a pattern of individual preferences into a pattern of group preferences.

If no further restriction were imposed, the arbitrator's job would be quite easy. He could, for instance, take the individual whose name comes first alphabetically and adopt his order of preferences for the group. This, obviously, is not in the spirit of what we are trying to do. Although the group order of preference may vary considerably—one person need not have the same influence as another, for example—some restrictions must be imposed. Arrow formulated several conditions which he thought a general voting scheme should satisfy. (See Appendix.) For one, the group preference should not be dependent solely on one player. In addition, given any two choices, *A* and *B*, there must be some pattern of individual preferences that will induce the arbitrator to say the group prefers *A* to *B*, and conversely. Working with these overall conditions, and other, more complicated ones, Arrow showed that it is impossible to formulate a general voting scheme. This being the case, we should not expect too much of any practical voting procedure.

Since a general solution to the arbitrator's problem does not exist (at least, not one that satisfies Arrow's conditions), we must settle for something else. We can try to construct a scheme, for instance, that is somewhat less general. In one special case that is particularly simple, all voters have "single-peaked" preference patterns.

Suppose there are a number of choices and they all fall

into some sort of order. If a budget is being prepared or a tax is going to be levied, the magnitudes would constitute such an order. If candidates are running for office, they may fall into order according to their positions in a left-right political spectrum. Each voter has some preferred place in the order, a preferred budget or political position. The voter's preferences are said to be "*single-peaked*" if between two alternatives on the same side of his first preference, A, he always prefers the one closest to A. If his first preference is a budget of $20,000, and two alternative budgets are both larger than $20,000, he will prefer the smaller one.

If every voter has single-peaked preferences, Duncan Black has shown, group preferences cannot be intransitive; there will always be one budget that will be preferred to all the others (barring ties). Black also showed that the preferred alternative is stable; that is, it is impossible for any subgroup of voters simultaneously to change their votes, thus effecting a change in the "preferred alternative" that all members of the subgroup simultaneously prefer. (In an earlier political example, we saw that masking one's true preferences by strategic voting can be advantageous, but not here.)

If we assume that in an election each of two candidates takes a position along a left-right spectrum and that the voters' preferences are all single-peaked with respect to this order, then Black's theory applies. If both parties are interested solely in winning the election and are unconcerned about ideological considerations (quite a simplification), they would take positions very close to each other. This would be true, whatever the composition of the electorate. (The composition of the electorate would influence where the common position is, however.) When one party takes a position and the other party has a choice of two positions on the same side as the first party, it always does better to take the one closer to the first party's.

The tendency of major parties to avoid extreme positions

and cluster about a middle point is well known to political observers. It is only a tendency, however, since ideological considerations *may* be significant and political opinions do *not* lie along a single dimension. More important, when the parties' positions are too close, the voters at the extremes will often vote as if their preferences were not single peaked. They may resent the fact that their vote is taken for granted while the established parties woo the middle-range voters, and they may refuse to vote "rationally." Instead, they may become what Anthony Downs calls "future-oriented," and vote for a hopeless third party, or abstain completely. They risk little thereby (there is little choice between the parties anyway) and obtain leverage on future elections, which pressures their party away from the center.

The problem may arise in an economic context also. Consider what happens when two competing companies decide to build a mill on a highway on which all their customers are located. Suppose both mills charge the same basic fee, plus a freight charge that is proportional to the customer's distance from the mill. If a customer buys from the mill that offers him the lowest price, then the decision the mills make in choosing a location is identical to the decision the political parties made. A. Smithies and Harold Hotelling predicted some time ago that the mills would be built close to each other (and near the median customer), for the same reason that political parties cluster about a center position.

If there is more than one serious candidate running for a single office, the voter's problem is more complicated still. Even if he is interested in the present election only and is willing to discount the effect of his vote on future elections, his strategy may not always be clear. Suppose, for example, the candidates are *A, B,* and *C,* and he prefers *C.* If he thinks it likely that his candidate will do worse than the other two and if the outcome is determined by a plurality vote, he may do better to give up on *C* and cast his vote for *A* or *B.*

In multi-candidate elections in which not one but many

offices are to be filled, and where the outcome is determined by proportional representation, the parties are not restricted to competing for the middle ground but may concede it to obtain a secure position at one extreme. An analogous situation was described in the *Harvard Business Review* by Alfred A. Kuehn and Ralph L. Day. In a study entitled "Strategy of Product Quality," they consider the case of a company manufacturing chocolate cakes, along with four other competing firms. If all five companies reach for that degree of "chocolaty-ness" most preferred by the median of the population, Kuehn and Day observe, each company can expect to get roughly a fifth of the market. If, instead of scrambling with the others for the middle ground, a company concentrates on one of the extremes, it has a chance of doing better.

Voting Model Applications

Because elections are of considerable concern to political scientists, various attempts have been made to construct models that will reflect the electoral process. The most interesting situation is one in which the voters can communicate easily and form coalitions—as in situation 4 in Buchanan and Tullock's hypothetical road-repair example.

In these games, political scientists are after much the same thing as the game-theorists: to determine which coalition will form and what the payoff will be. We have already mentioned one example of this type: the model of a political convention constructed by Polsby and Wildavsky. William Gamson formulated a theory for such games according to which a player's payoff is proportional to the resources he brings to a coalition. He tested the theory in the context of a mock political convention in which the players were "party bosses," the resources were the votes that the player controlled, and the payoffs were in terms of patronage: the jobs awarded. The number of jobs a player received, it turned out, was roughly proportional to the votes he controlled, as predicted.

In *An Economic Theory of Democracy*, Anthony Downs constructed a somewhat more ambitious model of the election process. He assumed that the electorate was rational and that its actions were "reasonably directed toward conscious goals." He also assumed that politicians have one basic goal—to stay in office—and that is what ultimately determines party policy. Specifically, Downs theorized, party leaders would endorse the policy that would maximize their party's vote.

On the basis of these assumptions and of some of the concepts we mentioned earlier (such as the space analogy of Smithies and Hotelling, the "future-oriented voter," etc.), Downs drew the following conclusions:

1. If, in a two-party system, a majority of the electorate are clearly on one side of an issue, both parties will be on that side of the issue, too.

2. In a two-party system, positions are less clearly defined and ideologies are vaguer than they are in a multiparty system.

3. A democratic government favors producers rather than consumers. That is, large, almost indifferent blocs have less influence than small, dedicated groups.

4. In a multiple-party election in which there is a single winner determined by plurality vote, smaller parties will tend to merge. Even if the electorate is represented proportionally, lesser parties will tend to merge, but to a lesser extent.

William H. Riker, in *The Theory of Political Coalitions*, rejects one of Downs's assumptions. He feels that a party will try to obtain a minimal winning coalition rather than capture the largest number of votes, since the greatest gain for coalition members comes when the majority is kept small. In situation 4 in Buchanan and Tullock's road-building problem, the larger the majority coalition, the smaller the advantage of the members. If there is a 99 percent majority in the coalition, for instance, a player receives almost what he would have obtained in situation 1.

Although there is some variation from theory to theory, an area of common agreement exists among political scientists. Most experts would agree, for instance, that public expenditures of public resources are not made in the most efficient or the most equitable manner. It is often expensive for a voter to acquire information that is relevant to his interests, and so he remains ignorant, especially if the issue involved does not affect him immediately. The interests of large numbers of partially informed, generally apathetic consumers are more than balanced by a small number of highly motivated, highly aware producers. On legislation concerning food and drugs, water purification, and safety control of aircraft, for instance, private interests are more powerful than their numbers might suggest. Even when the citizenry is informed, however, most government budgets are too small.

The A Priori "Value" of a Player

In March 1868, the directorate of the Erie Railroad, consisting of Drew, Gould, and Fisk, sought permission from the New York State Legislature to issue stock at will. They were opposed by Vanderbilt of the New York Central Railroad. The general approach required was well understood by the Erie directorate. According to one account, the members of the New York State Legislature "for the most part sold their votes at open bidding in the corridor of the State House." The only real question was how much should be set aside for bribes. If we assume that there are a fixed number of legislators who never abstain from voting, and that permission to issue stock at will has some definite dollar value to the Erie directorate, how much should the Erie directorate pay to bribe one member? Two members? n members? If the bill must be approved by another legislative body and an executive, how will this affect the answers?

L. S. Shapley and Martin Shubik found one answer based on what amounts to the Shapley value. (One can regard legislative bodies, executives, individual legislators, etc., as players in an n-person game; any coalition that has enough votes to pass a bill is called winning, and the others are called losing.) Shapley and Shubik concluded that the power of a coalition was *not* simply proportional to its size; a stockholder with 40 percent of the outstanding stock, for example, would actually have about two-thirds of the power if the other 60 percent of the stock was divided equally among the other six hundred stockholders. (When a person controls 51 percent of the vote, the other 49 percent is worthless if decisions are made by majority vote. Hence, power could not sensibly be defined as being proportional to the number of votes.)

The rationale on which Shapley and Shubik based their evaluation scheme is best illustrated by two examples:

Committee *A* consists of twenty people, nineteen members and a chairman. Committee *B* consists of twenty-one members, twenty members and a chairman. In both cases we will assume (for simplicity) that the members never abstain from the vote and that decisions are made by a simple majority. If the members are evenly divided on an issue, the tie-breaking vote is cast by the chairman. What is the power of the chairman in each of the two committees?

In the case of committee *A*, the answer is easy: the chairman has no power at all. The chairman votes only when the members split evenly, and nineteen members never split evenly. (There are no abstentions.) In committee *B*, the situation is a little more complicated. The chairman will cast a vote occasionally but will not vote as often as ordinary members. Does it follow that the chairman, who clearly has some power, is weaker than a member? Shapley thinks not. To see why, imagine that a proposition which the chairman favors is before the committee. There are three possibilities: the number of committee members who favor

the proposition is (1) less than ten, (2) ten, (3) greater
than ten. In 1, the chairman can't vote and the bill is de-
feated; but if the chairman could vote, *his vote would be
futile*. In 3, the chairman can't vote but the bill passes anyway; if the chairman could vote, *his vote would be re-
dundant*. Case 2 is the only case in which the chairman
can vote; it is also *the only case that matters*. Consequently,
Shapley and Shubik conclude, the chairman has the same
power as a member.

Using a similar but more general argument, Shapley and
Shubik show how an index of power can be derived. They
define the power of a coalition (or player) as that fraction
(of all possible voting sequences) of the time that the coalition casts the deciding vote; that is, the vote that first
guarantees passage. Thus, the power of a coalition is always between 0 and 1. A power of 0 means that a coalition
has no effect at all on whether a bill is passed; and a power
of 1 means that a coalition determines the outcome by its
vote. Also, the sum of the powers of all the players is always
equal to 1. If there are n players in a game and all votes
have the same significance, the power of each player is $1/n$,
as one would expect.

Shapley and Shubik also analyzed the power distributions
in more complicated situations. In the United States government, where it takes two-thirds of the House and Senate, or
one-half of the House and Senate plus the President, to pass
a bill, the entire Senate has the same power as the entire
House, and the President has about two-fifths the power
of either one. The President has about forty times the power
of an individual senator and about 175 times the power of
a member of the House. Shapley and Shubik gave another
example—which we mentioned earlier—of the well-known
phenomenon that large blocs of stock have more power
than the mere percentage of the total might indicate.

Some attempts have been made to apply the Shapley-
Shubik power indices to realistic problems. Typical of these
applications is one by Luce and Rogow. However, whereas

Shapley and Shubik derive their a priori power index from the abstract power structure exclusively and disregard such factors as tradition, ideology, personality, etc., Luce and Rogow have sought to incorporate into their analysis certain realities of Congress, such as special interests and geographic blocs. They start by making certain convenient but inessential assumptions to facilitate the analysis, such as that each member of Congress belongs to one of two parties; that there are no ties in the voting; that there are no abstentions; and so on. In addition, certain substantive assumptions are made that reflect specific relationships which are peculiar to the United States Congress—for instance, that there is not likely to be a coalition between Northern liberals and Southern conservatives on a civil-rights issue.

Luce and Rogow distinguish between regular party members and defectors. A member of a party is called a "defector" if he deviates from his party when he is so inclined. A senator or a member of the House may be a defector; so may the President. The power distribution in the Congress, it is assumed, is dependent on four factors: (1) the relative strengths of the two parties; (2) the number of defectors in each party; (3) which party is the President's; (4) the likelihood of the President's defection. Luce and Rogow analyze thirty-six possible variations of these four factors and include an illustrative example. Some well-known observations of political scientists are derived from the model. For instance, it is shown that if the President is from the majority party, the President will have more power if his party does not have two-thirds of the total vote.

In a study by Riker and Niemi, eighty-seven roll calls in the 86th Congress were analyzed with respect to twelve different issues, in the manner of Shapley and Shubik. If d members were on the winning side of a vote, each was credited with $1/d$ units, and the other members were credited with nothing. (This scheme was somewhat modified to account for the occasional absence of some of the members.) The conclusion was that there is, roughly, a set of blocs in the Congress, and these blocs have fairly constant power indices on some issues, though not on others.

Riker, in an earlier paper, suggested another way of "testing" the validity of the Shapley-Shubik power index. For this purpose, a legislative body with the following characteristics would be selected: (1) There would be tight party discipline, so that it would be reasonable to treat a party as a single person. (2) There should be more than two parties. (3) Members should migrate frequently from one party to another.

The French National Assembly seemed admirably suited to Riker's purpose. During one meeting, from 1953 to 1954 (the period of this study), there were between fifteen and twenty parties, eight of which had more than one member. Also, there was very tight discipline, and a 15 percent migration, i.e., 15 percent of the members changed parties in the course of this one session.

Riker attacked the problem as follows. He noted that the strength of a member was equal to the strength of his party divided by the number of members in the party. Thus, if there were three parties—A, B, and C, with 60, 70, and 80 members, respectively—each party would have a party index of one-third, since a party could not pass any legislation by itself but a coalition of any two of the parties could. The strength of an individual member of party A, B, and C would be $1/180$, $1/210$, and $1/240$, respectively. If the members of the French National Assembly perceived the power distribution in the same way that Shapley and Shubik did, the migration, it might be supposed, would be from those parties in which members had little power to parties in which members were stronger (disregarding ideological considerations, which presumably balance out). In our example, A would become larger, and C smaller. Though the results were not conclusive, this was apparently the case. (Strictly speaking, this is not really a test of the Shapley-Shubik result. Shapley and Shubik do not purport to measure the actual power distribution but merely give what they feel is a proper a priori power index based solely on the characteristic function and *nothing else*. Also, this study, if anything, reveals how the members of the Assembly perceived the power distribution. Nevertheless these findings suggest that "power," in the sense we are using here, is meaningful to legislators.) Finally, Schubert applied the Shapley-Shubik power index to the Supreme Court. Schubert observed that, often, most of the judges polarize on the two extremes of an issue, and a few judges take the middle ground. When neither extreme position is supported by a majority of judges, the final decision is determined by the pivotal judges in the middle. In such a case, the middle-of-the-road judges have power that is disproportionate to their numbers. Schubert assumes that the payoff in this "game" is the privilege of writing the majority opinion and shows that there is a significant correlation between the power indices of Shapley-Shubik and the payoffs.

A Summary

In constructing a theory of *n*-person games, the game-theorist has a choice. He can focus on one aspect of the game and disregard others, or he can try to capture all the relevant features of the game in a single model. In either case, he pays a price.

A-M chose the first course; they assumed that they were dealing with competitive players who could communicate freely and who have simultaneous and instantaneous access to one another. In those situations in which their assumptions are justified the theory is convincing and, for each coalition structure, may be used to predict what the range of payoffs will be.

N-M's purpose was quite different. It was not their intention to predict what the outcome of a game would be. Their concept of a solution was to be sufficiently general to encompass all possible "standards of behavior" that might arise.

Shapley's approach was particularly appealing because of its simplicity. But here also there was no attempt to predict. His "values" are only averages and arrived at by a way many of the factors which actually determine the outcome of a game.

It has been our intention to make this book accessible to as wide an audience as possible. To this end we have not discussed applications that are excessively technical. In particular we have not mentioned applications that have been made to international trade, the use of national resources, collective bargaining, etc. For the same reason we have excluded those topics which are of primary interest to the mathematically sophisticated such as proofs of the existence of A-M solutions or the minimax theorem for two-person, zero-sum games. Despite these omissions it is hoped that some idea of what the problems are and what has been accomplished has been obtained.

Starting with a framework sufficiently broad to encompass economics, politics, international relations, and many other fields, game-theorists have developed techniques that permit one to determine what a "player" can expect from a game (on average), what the specific payoffs should be for a given coalition structure, what the characteristics are of imputation sets that reflect society's "standard of behavior," and which of these are stable. In view of the complexity of the problem, this is all anyone could reasonably expect.

Appendix

The Arrow Social Welfare Theorem

Suppose a society that consists of two or more individuals is to construct a preference ordering over three or more alternatives on the basis of the preferences of its members. Such a situation arises when a legislature fixes its agenda (when the legislators have varying opinions concerning the bills' relative importance) or when a company is about to make certain investments in accordance with the desires of its stockholders. What is wanted is a *social welfare function*: some way of translating every possible preference pattern of the individuals into a single order of preferences for the society. We assume that for any two alternatives each individual will prefer one to the other or be indifferent between them; that is, any two alternatives are comparable. In addition, the following restrictions are imposed.

Restriction 1 If, on the basis of the preferences of the individuals of which the society is composed, the social welfare function dictates that alternative P is preferable to alternative Q, and if some of the individuals' preferences are altered so P becomes more desirable than it was earlier but other preferences are left unaltered, the social welfare function must still dictate that P is preferable to Q for the altered preference pattern.

Restriction 2 Suppose the social welfare function imposes a certain preference ordering in P: some part of the complete set of alternatives. If the preferences of the individuals are changed with respect to the alternatives outside of P but remain as they were when both alternatives are in P, the social welfare function will continue to impose the same welfare ordering within P.

Restriction 3 For any two alternatives there must always

be *some* pattern of individual preferences which will make society prefer one to the other and conversely.

Restriction 4 It cannot happen that there is one individual and two alternatives where society prefers one or the other of the alternatives precisely when the individual does.

The rationale for these restrictions is to make the decisions of society correspond, at least roughly, with the desires of the individuals who constitute it. If, for example, society always prefer x to y whatever the individuals prefer or, alternately, if one individual dictates what society should prefer, this is no group decision.

Arrow proved that it is impossible to construct a social welfare function satisfying all of these conditions.

Bibliography

Allais, Maurice. "Le comportement de l'homme rationnel devant le risque: critiques des postulates et axiomes de l'ecole Americaine." *Econometrica*, XXI (1953), 503–546.

Allen, Layman E. "Games Bargaining: A Proposed Application of the Theory of Games to Collective Bargaining." *Yale Law Journal*, LXV (1956), 660–693.

Anscombe, F. J. "Applications of Statistical Methodology to Arms Control and Disarmament." In the *Final Report to the U.S. Arms Control and Disarmament Agency under Contract No. ACDA/ST–3*. Princeton, N.J.: Mathematica Inc., 1963.

Arkoff, A., and Vinacke, W. E. "An Experimental Study of Coalitions in the Triad." *The American Sociological Review*, XXII (1957), 406–414.

Arrow, K. J. *Social Choice and Individual Values*. Cowles Commission Monograph 12. New York: John Wiley and Sons, Inc., 1951.

Aumann, R. J., and Maschler, Michael. "The Bargaining Set for Cooperative Games." In M. Dresher, L. S. Shapley, and A. W. Tucker, eds., *Advances in Game Theory*. Annals of Mathematics Study 52. Princeton: Princeton University Press, 1964, pp. 443–476.

Aumann, R. J., and Peleg, B. "Von Neumann-Morgenstern Solutions to Cooperative Games without Side Payments." *Bulletin of the American Mathematical Society*, LXVI (1960), 173–179.

Becker, Gordon M., and De Groot, Morris H. "Stochastic Models of Choice Behavior." *Behavioral Science*, VIII (1963), 41–55.

Becker, Gordon M.; De Groot, Morris H.; and Marchak, Jacob. "An Experimental Study of Some Stochastic Models for Wagers." *Behavioral Science*, VIII (1963), 199–202.

Berkovitz, L. D., and Dresher, Melvin. "A Game-Theory Analysis of Tactical Air War." *Operations Research*, VII (1959), 599–620.

———. "Allocation of Two Types of Aircraft in Tactical Air War: A Game-Theoretic Analysis." *Operations Research*, VIII (1960), 694–706.

Bixenstine, V. Edward; Polash, Herbert M.; and Wilson, Kellogg V. "Effects of Level of Cooperative Choice by the Other Player on Choices in a Prisoner's Dilemma Game." *The Journal of Abnormal and Social Psychology*, "Part I," LXVI (1963), 308–313. "Part II," LXVII (1963), 139–147.

Black, Duncan. "The Decision of a Committee Using a Special Majority." *Econometrica*, XVI (1948), 245–261.

———. *The Theory of Committees and Elections*. Cambridge: Cambridge University Press, 1958.

Bond, John R., and Vinacke, Edgar W. "Coalitions in Mixed-Sex Triads." *Sociometry*, XXIV (1961), 61–81.

Brayer, Richard. "An Experimental Analysis of Some Variables of Minimax Theory." *Behavioral Science*, IX (1964), 33–44.

Buchanan, James M. "Simple Majority Voting, Game Theory and Resource Use." *The Canadian Journal of Economic and Political Science*, XXVII (1961), 337–348.

Buchanan, James M., and Tullock, Gordon. *The Calculus of Consent*. Ann Arbor: The University of Michigan Press, 1962.

Caplow, Theodore. "A Theory of Coalition in the Triad." *The American Sociological Review*, XXI (1956), 489–493.

———. "Further Developments of a Theory of Coalitions in the Triad." *The American Journal of Sociology*, LXVI (1959), 488–493.

Cassady, Ralph, Jr. "Taxicab Rate War: Counterpart of International Conflict." *Journal of Conflict Resolution*, I (1957), 364–368.

————. "Price Warfare in Business Competition: A Study of Abnormal Competitive Behavior." *Occasional Paper No. 11.* The Graduate School of Business Administration, Michigan State University.

Caywood, T. E., and Thomas, C. J. "Applications of Game Theory in Fighter Versus Bomber Conflict." *Operations Research Society of America,* III (1955), 402–411.

Chacko, George K. *International Trade Aspects of Indian Burlap.* New York: Bookman Associates, Div. of Twayne Publishers, 1961.

————. "Bargaining Strategy in a Production and Distribution Problem." *Operations Research,* IX (1961), 811–827.

Chaney, Marilyn V., and Vinacke, Edgar. "Achievement and Nurturance in Triads in Varying Power Distributions." *The Journal of Abnormal and Social Psychology,* LX (1960), 175–181.

Coombs, C. H., and Pruitt, D. G. "Components of Risk in Decision Making: Probability and Variance Preferences." *The Journal of Experimental Psychology,* LX (1960), 265–277.

Crane, Robert C. "The Place of Scientific Techniques in Mergers and Acquisitions." *The Controller,* XXIX (1961), 326–342.

Dahl, Robert A. "The Concept of Power." *Behavioral Science,* XXII (1957), 201–215.

Davis, John Marcell. "The Transitivity of Preferences." *Behavioral Science,* III (1958), 26–33.

Davis, Morton. "A Bargaining Procedure Leading to the Shapley Value." *Research Memorandum No. 61.* Princeton, N.J.: Econometric Research Program, Princeton University, 1963.

Davis, Morton, and Maschler, Michael. "Existence of Stable Payoff Configurations for Cooperative Games." *The Bulletin of the American Mathematical Society,* LXIX (1963), 106–108.

Day, Ralph L., and Kuehn, Alfred. "Strategy of Product Quality." *Harvard Business Review,* XL (1962), 100–110.

Deutsch, Karl W. "Game Theory and Politics: Some Problems of Application." *The Canadian Journal of Economics and Political Science,* CXX (1954), 76–83.

Deutsch, Morton. "The Effect of Motivational Orientation upon Trust and Suspicion." *Human Relations,* XIII (1960), 123–139.

————. "Trust, Trustworthiness, and the F-Scale." *The Journal of Abnormal and Social Psychology,* LXI (1960), 366–368.

————. "The Face of Bargaining." *Operations Research,* XIX (1961), 886–897.

Downs, Anthony. *An Economic Theory of Democracy.* New York: Harper and Brothers, Inc., 1957.

————. "Why the Government Budget Is Too Small in a Democracy." *World Politics,* XII (1960), 541–563.

Dresher, Melvin. *Games of Strategy: Theory and Applications.* Englewood Cliffs, N.J.: Prentice-Hall, 1961.

Dumett, Michael, and Farquharson, Robin. "Stability in Voting." *Econometrica,* XXIX (1961), 33–43.

Edwards, Ward. "The Theory of Decision Making." *Psychological Bulletin,* LI (1954), 380–417.

————. "Probability-Preference in Gambling." *The American Journal of Psychology,* LXVI (1953), 349–364.

————. "Probability-Preference among Bets with Different Expected Values." *The American Journal of Psychology,* LXVII (1954), 56–67.

————. "The Reliability of Probability Preferences." *The American Journal of Psychology,* LXVII (1954), 68–95.

————. "Variance Preferences in Gambling." *The American Journal of Psychology,* LXVII (1954), 441–452.

Fellner, William. *Competition among the Few.* New York: Alfred A. Knopf, 1949.

Flood, Merrill M. "Some Experimental Games." *Rand Memorandum RM–789–1,* 1952.

————. "Some Experimental Games." *Management Science*, V (1958), 5–26.
Fouraker, Lawrence E., and Siegel, Sidney. *Bargaining and Group Decision Making*. New York: McGraw-Hill, 1960.
————. *Bargaining Behavior*. New York: McGraw-Hill, 1963.
Friedman, Lawrence. "Game-Theory Models in the Allocations of Adveritsing Expenditures." *Operations Research*, VI (1958), 699–709.
Galbraith, John Kenneth. *American Capitalism—The Concept of Countervailing Power*. Boston: Houghton Mifflin Co., 1952.
Gale, David, and Stewart, F. M. "Infinite Games of Perfect Information." In II. W. Kuhn and A. W. Tucker, eds., *Contribution to the Theory of Games*. Princeton: Princeton University Press, 1953, II, 245–266.
Gamson, William A. "A Theory of Coalition Formation." *The American Sociological Review*, XXVI (1961), 373–382.
————. "An Experimental Test of a Theory of Coalition Formation." *The American Sociological Review*, XXVI (1961), 565–573.
Griesmer, James H., and Shubik, Martin. "Toward a Study of Bidding Processes: Some Constant-Sum Games." *The Naval Logistics Research Quarterly*, X (1963), 11–22.
————. "Toward a Study of Bidding Processes, Part II: Games with Capacity Limitations," *The Naval Logistics Research Quarterly*, X (1963), 151–173.
————. "Toward a Study of Bidding Processes, Part III: Some Special Models." *The Naval Logistics Research Quarterly*, X (1963), 199–217.
Griffith, R. M. "Odds-Adjustment by American Horse-Race Bettors." *The American Journal of Psychology*, LXII (1949), 290–294.
Harsanyi, John C. "Approaches to the Bargaining Problem before and after the Theory of Games: A Critical Discussion of Zethuen's, Hicks', and Nash's Theories." *Econometrica*, XXIV (1956), 144–157.
————. "Bargaining in Ignorance of the Opponent's Utility Function." *The Journal of Conflict Resolution*, VI (1962), 29–38.
————. "A Bargaining Model for the Cooperative n-Person Game." In A. W. Tucker and R. D. Luce, eds., *Contribution to the Theory of Games*. Princeton: Princeton University Press, 1959, IV, 325–355.
Haywood, O. C., Jr. "Military Decisions and Game Theory." *The Journal of the Operations Research Society of America*, II (1954), 365–385.
Hoffman, Paul T.; Festinger, Leon; and Douglas, Lawrence H. "Tendencies Toward Group Comparability in Competitive Bargaining." In R. M. Thrall, C. H. Coombs, and R. L. Davis, eds. *Decision Processes*. New York: John Wiley and Sons, Inc., 1954, pp. 231–253.
Hotelling, Harold. "Stability in Competition." *The Economics Journal*, XXXIX (1929), 41–57.
Iklé, Charles, and Leites, Nathan. "Political Negotiation as a Process of Modifying Utilities." *The Journal of Conflict Resolution*, VI (1962), 19–28.
Kalisch, G. K.; Milnor, John W.; Nash, John F.; and Nering, E. D. "Some Experimental Games." In R. M. Thrall, C. H. Coombs, and R. L. Davis, eds., *Decision Processes*. New York: John Wiley and Sons, Inc., 1954, pp. 301–327.
Kaplan, Morton A. "The Calculus of Nuclear Deterrence." *World Politics*, XI (1958–1959), 20–43.
Karlin, Samuel. *Mathematical Methods and Theory in Games, Programming, and Economics*, Vols. I, II. Reading, Mass.: Addison Wesley, 1959.
Kaufman, Herbert, and Becker, Gordon M. "The Empirical Determination of Game-Theoretical Strategies." *The Journal of Experimental Psychology*, LXI (1961), 462–468.
Kelley, H. H., and Arrowood, A. J. "Coalitions in the Triad: Critique and Experiment." *Sociometry*, XXIII (1960), 217–230.
Kuhn, H. W. "A Simplified Two-Person Poker." In H. W. Kuhn and A. W. Tucker, eds., *Contributions to the Theory of Games*. Princeton: Princeton University Press, 1950, I, 97–103.
Lacey, Oliver L., and Pate, James L. "An Empirical Study of Game Theory." *Psychological Reports*, VII (1960), 527–530.

Lave, Lester B. "An Empirical Description of the Prisoner's Dilemma Game." *Rand Memorandum* P–2091, 1960.
Lieberman, Bernhardt. "Human Behavior in a Strictly Determined 3x3 Matrix Game." *Behavioral Science,* IV (1960), 317–322.
Loomis, James L. "Communication, the Development of Trust, and Cooperative Behavior." *Human Relations,* XII (1959), 305–315.
Lucas, W. F., and Thrall, R. M. "N-Person Games in Partition Function Form." *Naval Logistics Research Quarterly,* X (1963), 281–298.
Luce, Duncan R., and Raiffa, Howard. *Games and Decisions.* New York: John Wiley and Sons, Inc., 1957.
Luce, Duncan R., and Rogow, Arnold. "A Game Theoretic Analysis of Congressional Power Distribution for a Stable Two-Party System." *Behavioral Science,* I (1956), 83–95.
Lutzker, Daniel R. "Internationalism as a Predictor of Cooperative Behavior." *The Journal of Conflict Resolution,* IV (1960), 426–430.
———. "Sex Role, Cooperation and Competition in a Two-Person, Non-Zero-Sum Game." *The Journal of Conflict Resolution,* V (1961), 366–368.
McClintock, Charles C.; Harrison, Alberta; Strand, Susan; and Gallo, Phillip. "Internationalism-Isolationism, Strategy of the Other Player, and Two-Person Game Behavior." *The Journal of Abnormal and Social Psychology,* LXVII (1963), 631–636.
McGlothlin, William H. "Stability of Choices among Certain Alternatives." *The American Journal of Psychology,* LXIX (1956), 604–615.
Markowitz, H. "The Utility of Wealth." In *Mathematical Models of Human Behavior* (a symposium ed. by Jack W. Dunlap). Stanford, Conn.: Dunlap Assoc., 1955, pp. 54–62.
Maschler, Michael. "A Price Leadership Solution to the Inspection Procedure in a Non-Constant-Sum Model of a Test-Ban Treaty." In the *Final Report to the U.S. Arms Control and Disarmament Agency under Contract No. ACDA ST–3.* Princeton, N.J.: Mathematica Inc., 1963.
May, Mark A., and Doob, Leonard W. "Competition and Cooperation." *Social Science Research Council Bulletin No. 25* (1937).
van der Meer, H. C. "Decision-Making: The Influence of Probability Preference, Variance Preference and Expected Value on Strategy in Gambling." *Acta Psychologica,* XXI (1963), 231–259.
Mills, Harlan D. "A Study in Promotional Competition." In Frank M. Bass, ed., *Mathematical Models and Methods in Marketing.* Homewood, Ill.: Richard D. Irwin, Inc., 1961, pp. 271–301.
Miyasawa, K. "The N-Person Bargaining Game." *Econometric Research,* Group Research Memorandum No. 25 (1961).
Moglower, Sidney. "A Game Theory Model for Agricultural Crop Selection." *Econometrica,* XXX (1962), 253–266.
Morin, Robert E. "Strategies in Games with Saddle Points." *Psychological Reports,* VII (1960), 479–485.
Mosteller, Frederick, and Nogee, Philip. "An Experimental Measure of Utility." *The Journal of Political Economy,* LIX (1951), 371–404.
Munson, Robert F. "Decision-Making in an Actual Gaming Situation." *The American Journal of Psychology,* LXXV (1962), 640–643.
Mycelski, Jan. "Continuous Games with Perfect Information." In H. W. Kuhn and A. W. Tucker, eds., *Advances in Game Theory.* Princeton: Princeton University Press, 1964, pp. 103–112.
Nash, John F. "The Bargaining Problem." *Econometrica,* XVIII (1950), 155–162.
———. "Two-Person, Cooperative Games." *Econometrica,* XXI (1953), 128–140.
von Neumann, John, and Morgenstern, Oskar. *The Theory of Games and Economic Behavior.* Princeton: Princeton University Press, 1953.
Newman, Donald J. "A Model for Real Poker." *Operations Research,* VII (1959), 557–560.
Niemi, P., and Riker, William H. "The Stability of Coalitions on Roll Calls in

the House of Representatives." *The American Political Science Review*, LVI (1962), 58–65.
Polsby, Nelson W., and Wildavsky, Aaron B. "Uncertainty and Decision-Making at the National Conventions." In Polsby, Dentler, and Smith, eds., *Political and Social Life*. Boston: Houghton Mifflin Co., 1963, pp. 370–389.
———. *Presidential Elections: Strategies of American Electoral Politics*. New York: Charles Scribner's Sons, 1964.
Preston, Malcolm G., and Baratta, Phillip. "An Experimental Study of the Auction-Value of an Uncertain Outcome." *The American Journal of Psychology*, LXXI (1958), 183–193.
Raiffa, Howard. "Arbitration Schemes for Generalized Two-Person Games." Report M720–1 R–30 of The Engineering Research Institute, The University of Michigan, 1951.
Rapoport, Anatol. *Fights, Games and Debates*. Ann Arbor: University of Michigan Press, 1960.
Rapoport, Anatol, and Orwant, Carol. "Experimental Games: A Review." *Behavioral Science*, VII (1962), 1–37.
Read, Thornton. "Nuclear Tactics for Defending a Border." *World Politics*, XV (1963), 390–402.
Riker, William H. *The Theory of Political Coalitions*. New Haven: Yale University Press, 1962.
———. "A Test of the Adequacy of the Power Index." *Behavioral Science*, IV (1959), 120–131.
Robinson, Frank D. "The Advertising Budget." In S. George Walters, Max D. Snider, and Morris L. Sweet, eds., *Readings on Marketing*. Cincinnati: Southwestern Publishing, 1962.
Robinson, Julia. "An Iterative Method of Solving a Game." *Annals of Mathematics*, LIV (1951), 296–301.
Rose, Arnold M. "A Study of Irrational Judgments." *The Journal of Political Economy*, LXV (1957), 394–402.
Royden, Halsey I.; Suppes, Patrick; and Walsh, Karol. "A Model for the Experimental Measurement of the Utility of Gambling." *Behavioral Science*, IV (1959), 11–18.
Savage, Leonard J. *The Foundations of Statistics*. New York: John Wiley and Sons, Inc., 1954.
Schelling, Thomas C. "The Strategy of Conflict—Prospectus for a Reorientation of Game Theory." *The Journal of Conflict Resolution*, II (1958), 203–264.
———. "Bargaining, Communication and Limited War." *The Journal of Conflict Resolution*, I (1957), 19–36.
Schubert, Glendon A. *Quantitative Analysis for Judicial Behavior*. Glencoe, Ill.: Free Press, 1952.
———. *Constitutional Politics*. New York: Holt, Rinehart and Winston, Inc., 1960.
Scodel, Alvin. "Induced Collaboration in Some Non-Zero-Sum Games." *The Journal of Conflict Resolution*, VI (1962), 335–340.
———. "Probability Preferences and Expected Values." *The Journal of Psychology*, LVI (1963), 429–434.
Scodel, Alvin, and Minas, J. Sayer. "The Behavior of Prisoners in a 'Prisoner's Dilemma' Game." *The Journal of Psychology*, L (1960), 133–138.
Scodel, Alvin; Minas, J. Sayer; Marlowe, David; and Rawson, Harvey. "Some Descriptive Aspects of Two-Person, Non-Zero-Sum Games, II." *The Journal of Conflict Resolution*, IV (1960), 193–197.
Scodel, Alvin; Minas, J. Sayer; and Ratoosh, Philburn. "Some Personality Correlates of Decision Making under Conditions of Risk." *Behavioral Science*, IV (1959), 19–28.
Scodel, Alvin; Minas, J. Sayer; Ratoosh, Philburn; and Lipetz, Milton. "Some Descriptive Aspects of Two-Person, Non-Zero-Sum Games." *The Journal of Conflict Resolution*, III (1959), 114–119.
Shapley, L. S. "A Value for n-Person Games." In H. W. Kuhn and A. W.

Tucker, eds., *Contributions to the Theory of Games*. Princeton: Princeton University Press, 1953, II, 307–317.

Shapley, L. S., and Shubik, Martin. "A Method for Evaluating the Distribution of Power in a Committee System." *The American Political Science Review*, XLVIII (1954), 787–792.

Shubik, Martin. *Strategy and Market Structure*. New York: John Wiley and Sons, Inc., 1960.

Siegel, Sidney, and Harnett, D. L. "Bargaining Behavior: A Comparison between Mature Industrial Personnel and College Students." *Operations Research*, XII (1964), 334–343.

Simmel, Georg. *Conflict and the Web of Group Affiliations*. Translated by Kurt H. Wolff and Reinhard Bendix. Glencoe, Ill.: The Free Press, 1955. Originally published in 1908.

Simon, H. A. "Theories in Decision-Making in Economics and Behavioral Science." *The American Economic Review*, XLIX (1959), 253–283.

Smithies, A. "Optimum Location in Spatial Competition." *The Journal of Political Economy*, XLIX (1941), 423–429.

Snyder, Glen H. "Deterrence and Power." *The Journal of Conflict Resolution*, IV (1960), 163–178.

Stone, Jeremy. "An Experiment in Bargaining Games." *Econometrica*, XXVI (1958), 282–296.

Stryker, S., and Psathas, G. "Research on Coalitions in the Triad: Findings, Problems and Strategy." *Sociometry*, XXIII (1960), 217–230.

Suppes, Patrick, and Walsh, Karol. "A Non-Linear Model for the Experimental Measure of Utility." *Behavioral Science*, IV (1959), 204–211.

Thompson, G. L. "Bridge and Signaling." In H. W. Kuhn and A. W. Tucker, eds., *Contributions to the Theory of Games*. Princeton: Princeton University Press, 1953, II, 279–290.

Uesugi, Thomas J., and Vinacke, Edgar W. "Strategy in a Feminine Game." *Sociometry*, XXVI (1963), 75–88.

Vickrey, William. "Self-policing Properties of Certain Imputation Sets." In A. W. Tucker and R. D. Luce, eds., *Contributions to the Theory of Games*. Princeton: Princeton University Press, 1959, IV, 213–246.

Vinacke, Edgar W. "Sex Roles in a Three-Person Game." *Sociometry*, XXII (1959), 343–360.

Weinberg, Robert S. "An Analytic Approach to Advertising Expenditure Strategy." Originally published by the Association of National Advertisers, 1960. Reprinted in Robert S. Weinberg, "The Uses and Limitations of Mathematical Models for Market Planning." In Frank N. Bass et al., eds., *Mathematical Models and Methods in Marketing*. Homewood, Ill.: Richard D. Irwin, Inc., 1961, pp. 3–34.

Willis, Richard H., and Joseph, Myron L. "Bargaining Behavior I: 'Prominence' as a Predictor of the Outcomes of Games of Agreement." *The Journal of Conflict Resolution*, III (1959), 102–113.

Zermelo, E. "Uber eine Anwendung der Mengenlehre und der Theorie des Schacspiels," *Proceedings of the Fifth International Congress of Mathematicians*. Cambridge, II (1912), 501–504.

Index